EXTRACTING ACCOUNTABILITY

EXTRACTING ACCOUNTABILITY

Engineers and Corporate Social Responsibility

JESSICA M. SMITH

The MIT Press
Cambridge, Massachusetts
London, England

The MIT Press would like to thank the anonymous peer reviewers who provided comments on drafts of this book. The generous work of academic experts is essential for establishing the authority and quality of our publications. We acknowledge with gratitude the contributions of these otherwise uncredited readers.

This book was set in Adobe Garamond Pro and Berthold Akzidenz Grotesk by Westchester Publishing Services. Printed and bound in the United States of America.

Library of Congress Cataloging-in-Publication Data is available.

Names: Smith, Jessica M., author.
Title: Extracting accountability : engineers and corporate social
   responsibility / Jessica M. Smith.
Description: Cambridge, Massachusetts : The MIT Press, [2021] |
   Series: Engineering studies | Includes bibliographical references and index.
Identifiers: LCCN 2020052840 | ISBN 9780262542166 (paperback)
Subjects: LCSH: Engineering ethics. | Social responsibility of business.
Classification: LCC TA157 .S588 2021 | DDC 174/.962--dc23
LC record available at https://lccn.loc.gov/2020052840

10   9   8   7   6   5   4   3   2   1

For Lena Mae Rolston
May you find a career worthy of your insatiable imagination and your drive to make the world a more caring place.

Contents

Series Foreword

We live in highly engineered worlds. Engineers play crucial roles in the normative direction of localized knowledge and social orders. The Engineering Studies Series highlights the growing need to understand the situated commitments and practices of engineers and engineering. It asks, What is engineering for? What are engineers for?

Drawing from a diverse arena of research, teaching, and outreach, engineering studies raises awareness of how engineers imagine themselves in service to humanity and how their service ideals impact the defining and solving of problems with multiple ends and variable consequences. It does so by examining relationships among technical and nontechnical dimensions and how these relationships change over time and from place to place. Its researchers often are critical participants in the practices they study.

The Engineering Studies Series publishes research in historical, social, cultural, political, philosophical, rhetorical, and organizational studies of engineers and engineering, paying particular attention to normative directionality in engineering epistemologies, practices, identities, and outcomes. Areas of concern include engineering formation, engineering work, engineering design, equity in engineering (gender, racial, ethnic, class, geopolitical), and engineering service to society.

The Engineering Studies Series thus pursues three related missions: (1) advance understanding of engineers, engineering, and outcomes of engineering work; (2) help build and serve communities of researchers and learners in engineering studies; and (3) link scholarly work in engineering studies to broader discussions and debates about engineering education, research, practice, policy, and representation.

Jessica M. Smith's *Extracting Accountability: Engineers and Corporate Social Responsibility* shows that understanding the accountability of technoscientific corporations requires critically analyzing the agencies of the people who work within them. Observers regularly demonize the extractive industries and portray corporate social responsibility as an insidious example of capitalist deception. In a revealing ethnographic and archival study of engineers in the extractive industries, Smith critically examines actions and commitments that these accounts dismiss or ignore. She shows how engineers have framed corporate responsibility as an extension of the material and service benefits of engineering, enacting what she calls an "ethic of material provisioning." In so doing, engineers practice accountability to multiple publics, from the people who live closest to extractive operations to activists who oppose their industries. Smith concludes by demonstrating that one way to alter the accountability of technoscientific corporations is to alter the agencies of engineers who work in them.

—Gary Downey and Matthew Wisnioski, Series Editors

Prologue

Our lives shape the research that we do, whether we make those connections clear or leave only traces of them in our writing.[1] The practice of ethnography brings these interconnections into sharp relief. Through interviews, conversations, and the everyday sharing of life we gloss as "participant observation," our lives become entangled with those we seek to understand. As anthropologists Janet Carsten, Sophie Day, and Charles Stafford write, biography is a "part of the process of ethnography rather than separate from or prior to it."[2] It is not just that our life experiences lead us to particular research projects, to asking some questions instead of others. Our interactions with the people we study affect our own lives, and the connections we create "inform the moral judgements and ethical practices that pervade the experience of fieldwork."[3] This intermingling of lives and the emotional resonances it sets in motion are often viewed with suspicion, evident in accusations that anthropologists have become "too" close with their subjects. Closeness, this line of reasoning goes, sullies our ability to take a normative stance in relation to the lives we observe and participate in.

The kind of research I do requires a balancing act between "ethnography as an exercise in human empathy and anthropology as an exercise in cultural critique."[4] Ethnography usually hinges on intimacy and trust that cultural critique potentially damages. This is especially fraught space for those of us who recognize that our writing will likely end up being read by those who we write about. Diana Forsythe insightfully reflected upon her ethnographic research with technoscientists, writing, "Those of us who write about well-educated people in the United States can be sure that our informants will be able to read everything we publish. We can also be sure that they will not agree with everything we write."[5] Nancy Scheper-Hughes's raw account of being expelled from the Irish village where she and her family once lived, after residents took offense to her anthropological portrayal of them, has stayed in the back of my mind since the first time I read it.[6] The stakes of damaging relationships in one's field site are particularly high for those of us who do anthropology "at home" in some way. As Forsythe wrote, "Where home and field are contiguous or even identical, there is no 'elsewhere' for the fieldworker to return to."[7]

This book represents the second major research project in which I found myself confronting these dilemmas. My puzzlement at the internal and sometimes contradictory workings of corporations stretches back to my intertwined personal and professional trajectories. I grew up in a Wyoming town and family that revolved around mining. My father spent his career as a diesel mechanic for one of the world's largest coal companies, and both my sister and I worked as temporary laborers in that company's mines during summer breaks from college. I still remember the ironic ways that my coworkers commented on and managed their relationship with the company and its subsidiaries. They made fun of corporate discourses that bade them to practice "good teamwork," underlining the power differentials that distinguished some members of the team (technicians) from others (their supervisors). But they also proudly wore coats, hats, and belt buckles emblazoned with the company logo. A good portion of their retirement savings was invested in company stock. And they defended their companies and industry against criticism from others, both real and imagined. I later returned to that mining town as an anthropologist to conduct fieldwork, which provided the platform for writing an ethnography that examined how gender, kinship, and labor dynamics of the region had led to an unusually successful integration of women into the mining workforce there.[8] Engineers were present in that research project, but they were not the focus of it.

When I joined the faculty of the Colorado School of Mines in 2012, I found myself immersed in an academic institution whose faculty and students were almost exclusively dedicated to engineering and applied science, with a long-standing focus on the mining and petroleum industries. I felt bewildered by my new surroundings, from the students' lock-step progression through their major's course flowchart—I had gone to a liberal arts college, after all!—to the curious ways they vehemently distinguished engineering from "emotions."

In a very real sense, the past eight years of learning to work at Mines has felt like fieldwork. Through interactions with students and faculty, I began realizing that how our students are taught to conceptualize and solve problems had a lot to do with the challenges and frustrations I observed firsthand between engineers and technicians back home in the mines, which piqued my curiosity in engineering education. That interest grew as I soon found myself collaborating with faculty across campus to make visible the inherent social and political dimensions of engineering, as my arrival coincided with the fracking boom and the rising concerns of faculty and students to understand and address the growing firestorm it had ignited.

As I started getting to know Mines students and alumni, I realized that for them to think about their accountabilities to the public, they had to first make sense of the corporate context of their work. The miners I had come to know in Wyoming identified as part of the companies employing them, but they also had clear institutional space to separate their sense of self from the companies employing them—they weren't paid to care about the company on their days off, they frequently joked. Engineers, in contrast, occupied management and executive roles that seemed to demand such care. I quickly became captivated by scholarly debates about how and why the engineering profession in the United States had become entangled with managerial pathways inside of corporations.

Writing about engineers in the mining and oil and gas industries presented different sorts of challenges than did my first research project, which directly involved family and close friends. Studying these industries via engineers implicated my institutional home: the place where I showed up to work; the place where I taught students who held a variety of hopes, fears, and desires for corporate careers; the place where I socialized with engineering and applied science professors; and the place where I was seeking tenure and promotion, from an academic administration and university committee composed almost exclusively of engineers.[9] Mines was a very different institutional context than the anthropology departments where others in my field wrote treatises about technical professionals while remaining at arm's length from them. Echoing one of my interlocutors who struggled more with the internal than the external mining company politics, I had to *work with* the people I was supposed to be critiquing. I decided to make my intellectual project one of what Gary Downey— engineer, anthropologist, and science and technology studies (STS) and engineering studies scholar—calls critical participation. I planned and conducted my research already with an eye to it circulating at Mines and other engineering schools. I envisioned different strategies for my research and teaching to open up the questions Downey poses for engineering studies in general: what are engineers and engineering for?[10]

My own imbrication in the fields of practice I was studying has raised eyebrows among scholars who look suspiciously on those of us who get "too close" to our research. There is no question that my institutional location and biography shaped my research. In turn, the research shaped my institutional location and biography, prompting me to embrace engineering studies and engage in collaborative projects of curricular transformation. I conceptualized a successful National Science Foundation grant proposal for a research and teaching project that would (1) ethnographically develop a critical analysis of the intersection of engineering and corporate social responsibility (CSR) in the mining and oil and gas industries and (2) integrate more critical social scientific take on CSR inside the engineering curriculum at Mines and other schools with large mining and petroleum programs.

My attempts to cultivate more robust approaches to thinking about the accountabilities of corporations inside engineering education could be critiqued for potentially shoring up the moral authority of corporations. The kinds of questions we asked of and with our students exceeded the ethical possibilities of dominant CSR discourses, but it is also true that the kinds of critical self-reflection on industry practice we nurtured are foundational to the moral register of CSR in general. Although I wish to highlight and problematize my own positionality in relation to the research, I also caution against "purity politics" that presume that it would be possible for other academics to fully stand apart from the industries they critique.[11] As Alexis Shotwell writes in *Against Purity: Living Ethically in Compromised Times*, "Personal purity is simultaneously inadequate, impossible, and politically dangerous for shared projects of living on earth."[12] We are all complicit in corporate forms and in the mining and oil and gas industries in particular, though we occupy different positions in these networks and have different opportunities to shape them. We owe it to ourselves and to our others to do more than dutifully acknowledge the high carbon footprint of academic life and then launch into calls to simply do away with mining or fossil fuels or capitalism. The epilogue chronicles my own experiments in critical participation, so readers who are the most curious about the interweaving of my biography and the ethnography and engineering education efforts may wish to start reading there.

This book focuses on engineers who view social responsibility as central to their profession and their everyday work. This means that I have not presented an in-depth analysis of those who vociferously marginalized concerns about social responsibility—and they do exist. While I do not claim that the engineers profiled here represent their profession as a whole, there is much to learn from engineers who take public accountability seriously. We learn not by painting overly flattering portrayals of them to challenge dominant stereotypes of the profession but by giving them a good argument that is attentive to the complexity of their lives as they attempt to inhabit and detach from the corporate world. The ethnography proposes that the primary dilemma facing engineers is not a dearth of ethics that opens them up to becoming corporate automatons, as many would suspect. Rather, the primary dilemma is how to manage competing personal, professional, corporate, and public accountabilities as they attempt to craft themselves as ethical actors, to orchestrate a dense network of distributed agencies, and to enact corporate forms that are responsive to different judgments of what the world is and what it could become.

Acknowledgments

I am pleased to acknowledge the funding, institutional legitimacy, and scholarly networks opened up by the grant I received from the National Science Foundation's Cultivating Cultures for Ethical STEM program (award 1540298). The grant also allowed this book to be published in an open access edition. The opinions, findings, and conclusions in this book are my own and do not necessarily reflect the views of the National Science Foundation. The NSF funding allowed Nicole Smith to serve as a postdoctoral fellow on the grant—thank you for the interviews you contributed to the project, for your help putting on the workshops, and for your participation in the engineering education work. The grant also funded an intrepid group of researchers who persevered to organize, clean up, and analyze thousands of undergraduate student survey responses: Greg Rulifson, Shurraya Denning, Cassidy Grady, Juliana Lucena, Christopher Spotts, and Courtney Stanton. I thank Linda Battalora, Elizabeth Holley, Rennie Kaunda, Carrie McClelland, Susan Peterson, and Emily Sarver for opening their classes to our collaborative engineering education research.

I am grateful for each of the people who generously shared their time and their networks to make the research possible. By far, my favorite part of this project was getting to know the people whose careers form the backbone of the chapters in this book. Turning personal connections and complex lives into ethnographic "data" that can circulate beyond the interview context is exhilarating, but also somewhat disheartening. I wish I could have written a book about each of you, but I hope you will find some resonance with what I was able to include.

Chuck Shultz was an early and passionate advocate for teaching engineers to think about social responsibility dimensions of their professional practice. Your time, networks, and financial support made it possible for us to create corporate social responsibility as a generative space for research, teaching, and fundraising at Mines. I thank you, Louanne, and your whole family for generously supporting our Humanitarian Engineering program and students. I am indebted to Art Biddle and Stan Dempsey, who generously shared their archives with me. Thank you for your candor. It would take another book or more to do justice to the prescient work you did forty years ago. Art took a deep interest in engineering education, going out of his way to ensure that future generations of engineers would approach engineering problems with humility and respect for the people and places they encountered in their work. Joey Tucker also stands out as an exemplar Mines alum who was at the forefront of integrating social responsibility into engineering practice. Thank you for supporting this work, our program, and our students.

This project would be unthinkable without the Colorado School of Mines as its home. Juan Lucena introduced me to engineering studies and STS. The first book he loaned me—Edwin T. Layton Jr.'s *Revolt of the Engineers: Social Responsibility and the American Engineering Profession*—was the start of a collaboration that has never been dull. Thank you for your camaraderie. Mines is home to leading scholars pushing to make the inherent social justice and social responsibility dimensions of engineering visible and valued inside the curriculum. I have learned a great deal working with Linda Battalora, Robin Bullock, Stephanie Claussen, Sarah Hitt, Terri Hogue and her research group, Katie Johnson, Toni Lefton, Jon Leydens, Juan Lucena, Carrie McClelland, Junko Munakata-Marr, Dean Nieusma, Beth Reddy, Greg Rulifson, Jeff Shragge, Kate Smits, Ben Teschner, and Qin Zhu. Carl Mitcham's intellectual curiosity and drive will remain a source of inspiration for years to come. Kevin Moore provided crucial institutional support for our efforts to transform engineering education. Thank you to Priscilla Nelson and Erdal Ozkan for facilitating my collaborations with your faculty and students. I thank my colleagues in the Engineering, Design, and Society Division, who make our university an exciting and enriching place to be a scholar and teacher. My research on fracking in Colorado was made possible by the ConocoPhillips Center for a Sustainable WE²ST (Water-Energy Education, Science and Technology), which funded a remarkable student research group: Shurraya Denning, Frances Marlin, Austin Shaffer, Tom Van Ierland, and Skylar Zilliox. The book also benefited from years of teaching the Corporate Social Responsibility course at Mines, and I thank my students for our spirited discussions.

While Mines is an exciting place to teach and do research, writing the book required a sabbatical. I am grateful to Mines for the time away and the opportunity to rediscover the joy of reading and writing ethnography. Juan Lucena and Julia Roos took care of our Humanitarian Engineering program while I was away. I am thankful for a visiting international fellowship from the British Academy (VF1101988), which allowed me to spend a semester as a visiting scholar at the Department of Social Anthropology at the University of St. Andrews. The book benefitted greatly from the department's exciting intellectual environment and the chance to reimmerse myself in anthropology. I thank Mette High for the warm welcome in Scotland and a long, fruitful, and fun collaboration on energy and ethics. I am grateful to my daughter, Lena, and her dad, Mike, for their understanding and patience for the schedule adjustments required by the time away.

Gary Downey and Matt Wisnioski were a dream team of editors, especially by providing insightful advice on structuring the book, clarifying its contributions, and enhancing its appeal to readers. I especially enjoyed being able discuss the book with Gary and his graduate STS students at Virginia Tech. Dean Nieusma provided insightful feedback on the entire manuscript at a crucial juncture in the writing process. Thank you for helping me clarify its relevance to audiences beyond those I had initially imagined. Stuart Kirsch provided characteristically insightful and thoughtful feedback. Anonymous reviewers of the overly expansive manuscript provided generative critique that strengthened the book immensely. Katie Helke shepherded the manuscript through the publication process with great care.

I had the great fortune to share parts of the book in progress with scholars at the University of St. Andrews, the London School of Economics, Drexel University, Macalester College, Rowan University, Cal Poly–San Luis Obispo, Western Colorado University, and the University of Toronto

Troost Institute for Leadership Education in Engineering. I thank Mette High, Gisa Weszkalnys, Vincent Duclos, Amy Slaton, Gwen Ottinger, Roopali Phadke, Jordan Howell, Abel Chavez, and Doug Reeve for those visits. Workshops were particularly invigorating: the Rethinking Responsibilities colloquium at Harvard University, hosted by Suzana Trnka and Catherine Trundle; the Engaging Resources: New Anthropological Perspectives on Natural Resource Environments workshop at Wilfrid Laurier University in Waterloo, hosted by Gisa Weszkalnys and Tanya Richardson; the Energy Ethics conference at the University of St. Andrews, hosted by Mette High and myself; and the New Directions in the Energy Humanities workshop at Yale University, hosted by Doug Rogers and Paul Sabin. Sections of the book also benefited from discussion at conferences hosted by the Society for Social Studies of Science; the American Anthropological Association; the Society for Mining, Metallurgy, and Exploration; the Society for Petroleum Engineers; the American Society for Engineering Education; the National Humanities Conference; and the International Communication Association. Mike Rolston and my parents, Mike and Juanita Smith, graciously made it possible for me to travel to each of those lectures and conferences.

I drew great sustenance from my professional networks. I am grateful for the unshakable solidarity from the Liberal Education/Engineering and Society Division of the American Society for Engineering Education. Abby Kinchy, Roopali Phadke, and our STS Underground network have provided an exciting intellectual environment to think about the underground. I appreciate the wise advice of Gwen Ottinger, Beth Reddy, Christy Spackman, and Caitlin Wylie in our ad hoc committee on navigating academic life. I will be forever grateful for the University of Michigan intellectual community and friendships that continue to endure long beyond my time in Ann Arbor. Thank you especially to Kelly Fayard, Bridget Guarasci, Jessica Robbins, Mikaela Rogozen-Soltar, and Cecilia Tomori.

The final substantive revisions to this book were completed in the midst of the global COVID-19 pandemic and the protests calling for racial justice. Both of these events make it impossible to ignore the inherent social, environmental, and health injustices built into systems of scientific and engineering knowledge and practice. These massive upheavals were also a reminder of the dense social networks that sustain our intermingled personal and professional lives. I could not have found the time, space, and emotional presence to write without Lena Mae Rolston; Mike and Juanita Smith; George and Marilyn Smith; Luella Johnsen; Katie, Josh, Kaden, and Cameron Christy; Erin Roosa Cohen; and Kelly Fayard. When Lena was not with me during the unfolding crises, she had the good fortune to be with her dad; her grandparents; Ashley, George, Jack, and Luke Athanasopoulos; and Taylor Worsham.

All shortcomings in this book are my own.

Some of the material on the ethic of material provisioning in chapter 1 was previously published in the journal *Extractive Industries and Society*, some examples from chapter 5 were published in the journal *Engaging Science and Technology* in a paper I co-authored with Nicole Smith, and the epilogue's summary of the engineering education research was adapted from paper "Counteracting the Social Responsibility Slump? Assessing Changes in Student Knowledge and Attitudes in Mining, Petroleum, and Electrical Engineering," which I presented at the 2020 American Society for Engineering Education Virtual Annual Conference and that was co-authored by Greg Rulifson, Courtney Stanton, Nicole Smith, Linda Battalora, Emily Sarver, Carrie McClelland, Rennie Kaunda, and Elizabeth Holley. I thank the journals for the permission to repurpose that material here.

List of Figures

1 INTRODUCTION

When petroleum engineer Aaron set out to change how people in the oil and gas industry engaged the public, he described himself and his like-minded colleagues as "waging two wars": one inside their companies and one outside them. It was 2012, in the heady days of the oil and gas boom that brought energy production to places unaccustomed to it, such as the rapidly expanding suburban neighborhoods along the metro Denver Front Range in Colorado where Aaron worked (figure 1.1). The twinned technologies of horizontal drilling and hydraulic fracturing had sparked a rush for new drilling and intense controversy. A few Colorado cities and counties had approved bans and moratoriums on this kind of oil and gas development—which critics glossed as "fracking"—and companies like the one Aaron worked for scrambled to respond. Seeking to gain a better understanding of people's perceptions of their industry, Aaron and some of his coworkers began attending public meetings and hearings, where they described feeling attacked by residents and anti-fracking activists who accused them of endangering the health, safety, and livelihoods of their communities. They then returned to their company and the people whom they thought should have been their allies. "But instead of receiving a hero's welcome, we were treated as sympathizing with the enemy," Aaron recalled. "It was as if by *listening* to 'those people' we were sympathizing with them."[1]

Compelled to change the antagonism on both sides, Aaron continued looking for different ways to engage the public. While studying for an MBA he learned about the intense controversies the mining industry faced in the 1990s as it expanded globally. He described experiencing a lightbulb

Figure 1.1
A workover rig in close proximity to a Colorado neighborhood. Photo by milehightraveler, used by permission.

moment when he realized that the long-term success of his company and the entire oil and gas industry rested on community acceptance, just as the major mining companies had learned. He began translating the tools and techniques developed by the mining industry to oil and gas development in Colorado. He and his coworkers went on to devise public engagement strategies that would eventually transform how oil and gas company personnel in Colorado interacted with an array of stakeholders, from nearby residents to government officials.[2] They formed a stakeholder engagement team and went door to door in each of the neighborhoods where their company planned development, talking with residents to understand their concerns and answer their questions. They hosted community meetings in which people could talk directly with employees in more informal and dialogic ways than was allowed by the public hearings mandated by government regulation. They established a hotline that was answered in person by a stakeholder engagement team member, who logged and categorized each call and tried to address each complaint or concern. They set up booths at county fairs, beer fests, and science outreach events, making company employees available to whoever stopped by and wanted to talk. By the time I was getting to know others who worked for Aaron's company, a few years after the stakeholder engagement team began their work, even the most hard-nosed engineers spoke about the importance of the social license to operate, if only because it minimized risk to their company's investments. Other oil and gas companies operating in Colorado began calling them to seek guidance in replicating their efforts, and similar public engagement techniques began appearing all over the Front Range (figure 1.2).

The experiences of engineers like Aaron who found themselves held accountable for issues of broad public concern provide a crucial window through which to analyze much broader concerns about the accountability of large, technoscientific organizations in the era of corporate social responsibility (CSR).[3] These organizations have tremendous potential to affect the well-being of the people who live near their operations or who make or

Figure 1.2
Education and outreach booth hosted by an oil and gas company. Employees invited children, such as my daughter featured here, to engage in hands-on learning about topographic maps and the stratigraphy of the formations where the company drilled to produce oil and gas.

use their products. Political theorist Langdon Winner influentially named engineers and technical professionals as the "unacknowledged legislators of our technological age," given the inherent politics of the technologies they design, build, and maintain.[4] Philosopher Carl Mitcham has noted that, "by designing and constructing new structures, processes, and products, [engineers] are influencing how we live as much as any laws enacted by politicians."[5] Yet corporations can shield themselves from public scrutiny, and asymmetrical access to knowledge and expertise further excludes many members of the public from participating in decision making about those structures, processes, and products.

The field of CSR arose in direct response to these and other critiques, positioning corporations as voluntarily promoting the well-being of people and the planet, in addition to profit. Like other forms of audit and accounting, CSR gains its power from the "twinned concepts of economic efficiency and good practice."[6] While corporations had long engaged in philanthropy, some in the 1960s and 1970s began focusing on specific issues such as urban decay, racial discrimination, environmental pollution, and workplace safety.[7] Stakeholder theory flourished in the 1980s, legitimizing efforts to engage people beyond shareholders and employees. In the wake of growing public pressure in the 1990s, many businesses began staking their reputation on their contributions to sustainable development, including in the mining and petroleum sectors.[8] In the 2000s, frameworks such as "natural capitalism," "shared value," and B Corps certification proposed that companies could be socially and environmentally responsible while being profitable.[9] The appeal and hazards of all these versions of CSR share much with historian Matthew Wisnioski's assessment of earlier invocations of reform in engineering: "The rhetoric of responsibility was infinitely malleable and universally desirable, which gave reformers legitimacy but also fostered cooptation."[10]

CSR as a field of practice is internally variegated. Though *corporate social responsibility* generally refers to the notion that corporations have obligations to society beyond generating profits, the term means different things depending on the industry, company, geographic or institutional location, and person invoking it. There was great interpretive flexibility even among my relatively focused group of interlocutors, who variously viewed CSR as philanthropy, sustainability, health and safety, and/or community relations and who practiced CSR as a rhetorical strategy, a management technique, a collection of performance standards, or an ethical goal. For some of them, CSR was separate from what they viewed as their "technical" work as engineers, while others viewed CSR as integral to their professional practice. Some instrumentally limited CSR to a useful tool to minimize social risk to investment, while others viewed it as a mandate to transform industry. The concept of CSR itself is thus best understood as a boundary object: a set of information that is interpreted and used in different ways by different people while maintaining a common identity. As proposed by their theorizers, boundary objects are "plastic enough to adapt to local needs and constraints of the several parties employing them, yet robust enough to maintain a common identity across sites."[11] Like other boundary objects, CSR facilitated collaboration without consensus, providing a common organizing framework for employees with otherwise disparate values, goals, and responsibilities.[12]

This vagueness leads critics to argue that "the use of the term CSR has become so broad as to allow for people to interpret and adopt it for many different purposes" and for businesses to "appropriate the meaning of ethics."[13] While even a cursory glance at CSR in practice shows that it is not a panacea for reconciling ethics and economics or morality and the market, nor can it be dismissed as disingenuous greenwash. The suite of concepts and practices that make up the field of CSR effect politics of their own. For example, the ubiquitous term *stakeholder* can reduce oppositional groups or government regulatory agencies to "simply individual players in a multi-stakeholder process that purports to accord equal footing to governments, corporations, communities, and civil society organizations."[14] In engineering practice, the stakeholder framework opens up a wider array of people to whom engineers ought to be accountable but leaves the power to define who legitimate stakeholders are, how they will be prioritized and engaged, and what engineers will do as a result of that engagement largely with engineers and other "systems owners."[15] Rather than generating consensus, CSR concepts and tools seed new forms of dispute, concerns, sites, problems, and subject positions for people who are employed by corporations, as well as those who seek to critique them.[16]

This book argues that, to understand the accountability of technoscientific corporations, we must understand the agencies of the people who constitute them. While some engineers I met depoliticized their work by restricting it to technical concerns, it was far more common for them to ground their work within larger moral aspirations of being "good" people or creating "good" in the world. The corporate context of their work, however, meant that they frequently found themselves trying to reconcile contradictions within and among multiple domains of accountability, including formal standards and policies, public demands, their personal ethical frameworks, and professional norms that enjoin engineers to serve as loyal agents for corporations while protecting the safety, health, and welfare of the public. Moreover, the corporate context of their work presented particular opportunities and constraints for those attempts at reconciliation: engineers experienced a distributed agency in which they were not always authors of their actions and frequently acted through others.

The multiple domains of accountability and distributed agency I observed are central features of engineers' corporate employment that likely stretch beyond my particular focus on the mining and oil and gas industries. But there are also specificities of these industries that raise caution for hastily extrapolating this book's findings to others. To start, mining and oil and gas companies deal with far more material stuff, wresting resources from the ground, than do those that trade in information. Moreover, the mining and oil and gas industries have a particularly thorny public reputation that distinctly colors how industry actors imagine and practice accountability to their others.

UNSTEADY MORAL TERRAIN

The mining and oil and gas industries have served as easy targets for social scientists and social movements committed to ideals of justice. All industrial activities have environmental and social effects. Many communities have found ways to coexist with or even welcome such activity, but the effects have been devastating for others, from attention-grabbing oil spills and dam failures to the "slow violence" of long-term pollution.[17] These harms have disproportionately been borne by poor and minoritized populations around the world, making the extractive industries a key engine—as well as product—of racial capitalism.[18] All too often, elites promote extractive activity as a means of national economic development while failing to safeguard the well-being of the people who live closest to such activities.[19] While I was researching and writing this book, two major mine tailings dams in Brazil failed, one in 2015 and one in 2019, claiming nearly three hundred lives while polluting hundreds of miles of waterways. In 2016, Native American activists on the Standing Rock Reservation galvanized a vibrant grassroots movement to protest the Dakota Access Pipeline, which they argued violated international human rights standards for informed consent, protection of sacred ground, and water quality. In 2020, an iron ore mine in Western Australia bulldozed—with legal permission—a sacred Aboriginal site that showed forty-six thousand years of continual occupation and provided a four-thousand-year-old genetic link to present-day traditional owners. Long-standing patterns of troubling appropriations of resources and wealth have led academics to identify the "extractive logics" of even renewable energy projects.[20]

The mining and oil and gas industries also come under fire as "angels of the Anthropocene."[21] In part, this is due to their contributions to anthropogenic climate change. But even more fundamentally, these industries are emblematic of the Anthropocene in how they move and transform massive amounts of earthen materials. Already by the late nineteenth century, copper mines in Butte, Montana, stretched a mile underground. If their inner infrastructure could have been turned on its head and stretched upward, it would have stood twice as tall as the world's tallest skyscrapers. When the state's Anaconda Smelter smokestack was completed in 1918, it was classified as the "largest freestanding masonry structure in the world," soaring thirty feet taller than the Washington Monument.[22] Mining equipment used around the world became progressively larger until, by 1980, "giant shovels, trains, and trucks moved more earth on the planet than did the forces of natural erosion."[23]

Current offshore oil production is emblematic of the increasingly complex—and risky—sociotechnical systems required to access ever more difficult resources.[24] The Perdido oil platform in the Gulf of Mexico, for example, is one of the world's deepest, floating in about eight thousand feet of water with an estimated peak production of one hundred thousand barrels of oil per day. It was built at a cost of $3 billion and is 267 meters tall, stretching almost to the height of the Eiffel Tower. The 2010 *Deepwater Horizon* disaster, the largest marine oil spill in the history of the petroleum industry, made clear the extent to which these kinds of technologies raise the stakes for errors and accidents.

Engineers play particular—but not omnipotent—roles in the sociotechnical systems that can facilitate or slow the sweeping global transformations glossed as the Anthropocene. Academics struggle to theorize accountability in these vast social, material, and environmental assemblages, generally taking one of two approaches that seem to be characterized by opposite impulses. On the one hand, "anti-anthropocentric" scholarship emphasizes materiality, objects, and other species, which may "index and oppose the toxic legacies of radically human-centered thinking and action."[25] This approach risks obscuring questions of human beings' differential responsibilities for the current predicaments in which we find ourselves. As Arjun Appadurai asks:

> If agency in all its forms is democratically distributed to all sorts of dividuals, some of which may temporarily be assembled as humans and others as machines, animals, or other quasi agents, then do we need to permanently bracket all forms of intrahuman judgment, accountability, and ethical discourse? . . . Will our very ideas of crime and punishment disappear into a bewildering landscape of actants, assemblages, and machines? If the only sociology left is the sociology of association, then will the only guilt left be guilt by association?[26]

On the other hand, academic and popular criticism can also take the extreme opposite approach by holding very particular actors responsible for the harms of entire systems. Implied in conference hallway jokes, podcast banter, and published research is the belief that the social and environmental harms wrought by natural resource production result from engineers and other corporate actors knowingly (and perhaps gleefully) placing profit above their other responsibilities.[27] This reading collapses the effects of people's everyday actions with their motivations: harms must be the result of a faulty or absent moral compass. My interlocutors picked up on this attribution of blame and universally described feeling demonized by people outside industry. One explained, "I tell people I'm a petroleum engineer and they just give me the stink eye. They think I'm evil. [They think] 'Oh, earth raper,' right?" Others refused to tell their neighbors and social circles that they worked in industry because they had experienced such severe judgment by them (more on this in chapter 4). This individualization of blame is likely due at least in part to legal systems that emphasize intentionality and locate culpability with individuals. But it might also be rooted in a more fundamental American moral imagination of nature. Religious studies scholar Evan Berry argues that Christian understandings of individual salvation underlie dominant American imaginations of nature and hamper our ability to imagine and grapple with the collective challenges of climate change.[28]

The engineers and applied scientists I came to know each explicitly or implicitly placed their work within their larger life projects of creating "good" in the world.[29] Like Andrea Ballestero's interlocutors, they considered "their technical work . . . a tool to attain ethical goals."[30] These moral ambitions and desires to craft themselves as ethical professionals thus framed how they thought about and practiced accountability. They acknowledged and grieved mistakes—both personal and collective—that had caused harm for people and environments. But even the most hardnosed engineers who denigrated CSR for detracting from profits still imagined themselves as morally righteous actors, for example, by putting their faith in corporations to create wider societal benefits or by justifying their actions within the ethic of material provisioning (see chapter 2).[31] One of the questions this book poses, therefore, is *how a group of professionals who all believe they are doing the "right thing" end up facilitating industrial development that can be judged as ethically suspect by others—and, at times, by themselves.*

Hannah Appel's ethnography of US oil companies in Equatorial Guinea sheds some light on this question. Theorizing the "licit life of capitalism," she shows how exploitation is made through devices that are "legally sanctioned, widely replicated, and ordinary": contracts and subcontracts, infrastructures, economic theories, corporate enclaves, and transparency projects.[32] In her analysis, these devices serve "not only as powerful tools in and of themselves, but also as a felicitous moral architecture through which to sanction capitalist practices."[33] This book complements Appel's ethnographic focus on these devices by deeply exploring how industry personnel themselves imagine and navigate the competing accountabilities that characterize corporate work. In this process they create, transform, reinforce, and undercut various and sometimes competing "moral architectures." Rather than creating a mutually reinforcing ethical framework that provided clear guidance for practice and decision making, I will show that corporate work generated multiple accountabilities that engineers and other technical professionals worked furiously to reconcile.

This ethnography shows that the "problem" of engineers' contribution to the extractive industries is not that they lack an ethical framework and therefore embody corporate drives for profit. Rather, the corporate context of their work generates competing accountabilities that engineers attempt to reconcile without clear guidance on how to do so. This analysis invites us to see how ethical dilemmas stem from multiple sources beyond the intentions or (in)actions of particular individuals. They stem from notions of accountability that frame questions of responsible extraction as *how* rather than *if*; from a division of labor and authority that disperses agency and responsibility among corporate engineers and consultants who all imagine they are doing the "right thing" but who pass recommendations and decisions off to others; from engineers who, when they reach their limits of being able to shape their colleagues' agencies to align more closely with their understanding of right action, reluctantly throw up their hands and distance themselves from the corporate person; and from engineers becoming so frustrated with the constraints of their work that they quit their corporate jobs, only to be replaced by a fresh batch of new hires.

STUDYING CORPORATIONS

Academic and public criticism often blackboxes corporations, treating them as unitary entities with an unwavering dedication to profit maximization. In contrast, understanding the accountability of technoscientific corporations by investigating the agencies of the people who constitute them is predicated on analytically disaggregating the corporate form.[34] Companies do exist as legal persons that have juridical agency, such as to make and break contracts. But this legal fiction of singularity belies the dense, uneven, and nested geographies of parent companies, subsidiaries, contractors, and subcontractors through which corporations actually act in the world. This "archipelagic" corporate form powerfully visualized by Appel has direct implications for its multiple accountabilities:

> This disaggregation or dispersion is, in effect, the legal (licit, intentional) thinning of liability, accountability, and responsibility, such that what seems clearly to be the singular exercise of corporate power—global companies in contract with governments around the world, maneuvering the world's largest mobile infrastructures and reaping spectacular profit—in practice fractures rapidly into a legally slippery tangle of subsidiaries and consortia and subcontractors.[35]

It is difficult to pin accountability to one node of a dizzying archipelago.

The legal fiction of corporate singularity also masks the internal heterogeneity of corporate forms. Anthropologists emphasize that corporate forms "act" in the world through the everyday practices of their employees.[36] This perspective opens up what can be perceived externally as a unitary corporate "person" to be a collective one, characterized by internal fissures and contradictions.[37] Marina Welker's ethnography of Newmont mining company's CSR activities shows that corporations must be made to "hang together"[38] in the face of competing enactments and contestations over a corporation's boundaries, interests, and responsibilities.[39] In her analysis, corporations are enduring and powerful because they are partible, composite, permeable, and in flux.

Tracking movement within corporate forms underscores Welker's point. The engineers I met experienced a "work organization in perpetual flux, with teams forming and disbanding, and team members and supervisors constantly circulating around the country and, indeed, all over the globe."[40] This constant movement makes it difficult to hold corporate forms accountable for the promises made by particular personnel, who move from site to site. This slipperiness is compounded when projects or entire companies are bought out by other companies that bring their own personnel and policies to bear on their operations and external engagements. In fact, one of the great sources of harm in these industries is that, while particular personnel and entire companies can leave, the industrial site remains for the people to whom promises were made and then broken.[41]

Some social scientists and other social critics have been wary of disaggregating the corporate form, preferring instead to portray them as homogeneous actors motivated by profit maximization—*homo economicus* in institutional form. This may be because more distributed understandings of corporate personhood make it difficult to hold them accountable for harm.[42] But disaggregating the corporation is a crucial step for analyzing "the messiness and hard work involved in making, translating, suturing, converting, and linking diverse capitalist projects . . . that enable capitalism to appear totalizing and coherent."[43] Feminist anthropologists show that the social relations of what we call "capitalism" are generated out of divergent life projects, not inexorable logics.[44]

Thus rather than presume that "capitalism" drives engineers to privilege profit at whatever social and environmental cost, this book investigates how engineers' invocations of shareholder value or appeals to the business case are intertwined with their efforts to craft themselves as ethical persons, chiefly by reconciling personal, professional, and corporate accountabilities. Engineers are key, if often overlooked, actors whose work sustains corporate forms and, in the process, shapes their accountabilities to multiple publics.

WHY ENGINEERS?

As the philosopher Carl Mitcham writes, "Engineering is everywhere, but not everywhere recognized."[45] Engineers help set mining and oil and gas activity into motion, and the reverberations of their decisions and designs—and the assumptions and desires built into them—echo beyond their own tenure at a company or even their own lifetime (figure 1.3).[46] Jon A. Leydens and Juan C. Lucena highlight the importance of engineers to questions of social justice, arguing that "engineers design, build, and operate complex and imposing systems, capable of influencing the lives of millions of people, as well as the allocation of resources (e.g., water, energy), opportunities (e.g., access to work and commerce), risks and harms (e.g., flooding, nuclear disasters, groundwater contamination), and how different social groups receive these differently."[47] One of the great paradoxes is that, although the infrastructures they design, build, and maintain exert a great influence the everyday lives and potential futures of people around the world, engineers themselves are also particularly situated actors, whose educational opportunities and work settings place constraints on what they learn, know, and do.

Engineers' place in anthropology is relatively small but growing, primarily due to increased interest in infrastructure.[48] The mining and oil and

Figure 1.3
Overlooking the Nevada Twin Creeks mine. Photo courtesy Nevada Bureau of Land Management.

gas industries, for example, are most often studied from the perspective of the people who organize to address their harms for vulnerable populations and environments.[49] Anthropological research that does include ethnographic attention to engineers and engineering vividly demonstrates the politics, exclusions, and harms embedded in infrastructure otherwise cloaked in the banners of neutrality and progress, from the politics of dams, pipelines, roads, and electricity markets to the flow—and lack thereof—of water.[50] Fabiana Li's research examines the role of engineering knowledge in mining-related controversies in Peru, including how the structural conditions of engineers' employment shape their ability to bring social and environmental concerns into their professional practice. She also shows that engineers and local campesinos differently understand phenomena such as water quality.[51] Martin Espig and Kim de Rijke call attention to the differences between how engineers and the people who live closest to coal seam gas production understand risk and uncertainty.[52] David Kneas shows how a junior mining company constructed geological assessments of copper mineralization in Ecuador to sell the "potential and possibility" of a copper resource to be mined, forming part of a much longer history of the contested creation of geological knowledge about the subsoil.[53] David McDermott Hughes also ethnographically demonstrates, through fieldwork in Trinidad and Tobago, how graphical representations construct oil resources and reserves.[54]

The relative dearth of "inside the fence" studies of these industries is partially due to the power of corporations to control access to production sites and headquarters. The social scientists who have been able to conduct research inside mining and oil companies or with their personnel have focused on externally-facing groups, generating rich research critical of CSR. Dinah Rajak's pioneering study of the mining multinational Anglo American documents and theorizes how CSR extends the moral authority of corporations. She found that CSR practitioners brought deeply held personal passions of "doing good" to their work of "empowering" the subjects of their programs but ultimately reinscribed coercive gift relationships with them that inspired "deference and dependence rather than autonomy and empowerment."[55] Welker's study of Newmont CSR personnel shows that they enacted the company to different ends—as a "pot of money" versus a "set of skills"—as they attempted to ameliorate the harms created by mining activities.[56] Douglas Rogers argues that the practice of CSR during the postsocialist oil boom in Russia's Perm region produced an "interpenetration of corporation and state" and remade the region through widespread cultural projects that played on the materiality of oil and gas and their attendant infrastructure.[57] John R. Owen and Deanna Kemp have conducted perhaps the most extensive research inside mining companies from their positions as researchers at the University of Queensland's Sustainable Minerals Institute. Working "inside the fence" allows them to show how efforts at sustainable community development can be undermined by grounding calls for social responsibility inside the business case for the social license to operate; to document how CSR practitioners experience marginalization inside corporate structures that leave them out of major decision making; and to identify voices for change inside companies that are "holding ground against the narrow business case view of the world."[58]

EMPATHY AND ETHNOGRAPHY

The focus on CSR and community relations personnel in these "inside the fence" social studies of mining and oil and gas industries reveals something deeper about the lack of attention to engineers in these literatures. The CSR and community relations personnel can be interpreted as valiantly trying to ameliorate or prevent the environmental and social effects of these industries.[59] This perceived quality may make them more appealing research subjects for social scientists, especially ethnographers. Ethnographic research methods invite empathy, fostering a "well-rehearsed disciplinary ideal of ethnographic encounters suffused with mutuality."[60] This mutuality may lead anthropologists to be wary of representing their interlocutors in a negative light, a phenomenon Sherry Ortner terms "ethnographic refusal."[61]

This ethnographic ideal of mutuality sits awkwardly beside the "hermeneutics of suspicion" demanded by social science in general and critical studies of capitalism in particular.[62] In this mode, we document the failures, the contradictions, and the betrayals of high ideals. The stark injustices engendered by late capitalism and climate change seem to demand such suspicion, especially when studying elites. When I was presenting my research to fellow academics, for example, many insinuated or outright asserted that my obligation as an anthropologist studying the mining and oil and gas industries was to pull back the curtain and expose the "true" ill intentions of industry personnel that were hidden by corporate greenwash. This approach ironically celebrates academic criticality though reconfirming common tropes about corporations and the people who enact them. As Welker writes:

> By making the profit-maximizing corporation the central protagonist, we perform our criticality in opposition to a corporate actor while disengaging "it" from the human and nonhuman agents involved in enacting and contesting corporations and their responsibilities. . . . The political satisfaction afforded by the performance comes at an ethnographic and epistemological cost, severing corporations from the ordinary materials, human practices, ethics, and sentiments (such as desire, fear, shame, pride, jealousy, and hope) that sustain them.[63]

Turning empathetic relationships and shared histories of encounter into "research" that travels beyond them and engages in cultural critique presents particular challenges for anthropologists who research highly politicized and polarizing topics. The few who study "up" may choose to forgo ideals of mutuality from the start.[64] Hughes, for example, vociferously argues that the unprecedented global threat of climate change merits judging petroleum engineers and scientists as "in the wrong." Because of their complicity in climate change, he proclaims, they and their industry should be consigned "to an ash heap, worthy of condescension and worse" and "should go extinct."[65] Most others choose to conduct research with the people and in the places that experience harm. In these cases, anthropologists can amplify, contextualize, and theorize their interlocutors' experiences to critique political, economic, cultural, and other structures of power alongside them.

In contrast to romantic ideals of mutuality, the structural position of many engineers as dutiful employees of major corporations and government agencies can make them, in the eyes of others, unsympathetic ethnographic subjects. Through their formal education and work experiences, engineers are socialized into professional worldviews that are animated by particular—if unstated—goals and assumptions about that world, what "progress" entails, and how it can be achieved. This has meant that in "most theorists' conceptions, engineers were the embodiment of the military-industrial complex: conformist organization men in the system that stood to be torn down."[66] Indeed, engineers' centrality to highly problematic development schemes, from high modernist state-led projects[67] to those seeking local community empowerment,[68] has seeded critiques of them as "hypersubjects"[69] who relentlessly pursue profit and efficiency. As Penny Harvey and Hannah Knox write in their ethnography of roads in Peru, "From such scenarios it is not difficult to see why the engineer, as modern expert, emerges as the villain in the critical social sciences."[70]

Such black-and-white ethical judgments hinder us from understanding how the very ethical commitments held by engineers nonetheless can help sustain the corporate forms they struggle within and against. The book follows María Puig de la Bellacasa in approaching the ethical as an everyday doing that "connects the personal to the collective" and grounds "ethical obligation in concrete relationalities in the making rather than on moral norms."[71] This approach does not necessitate a "postcritical" stance, justifying engineers' activities or evacuating our own moral and political commitments as researchers.[72] These ethical doings are inherently political, as they enable and are enabled by processes that unevenly distribute risks and benefits, harms and rewards.[73] As Ballestero cautions, technoscientific tools "quietly determine the limits of the possible by both narrowing down certain options and opening the possibility of creating different, and maybe better, worlds."[74] But understanding how people themselves judge the rightness and wrongness of the thoughts, activities, and relationships that make up their lives, Mette High and I argue, provides a crucial first step to better understanding our interlocutors—and the industries they help set into motion—so that we can then engage in more generative debate about possible futures.[75]

Taking seriously the everyday lives of engineers, for example, allows us (1) to trace how infrastructures, products, and processes come into being through those engineers' own moral projects and then (2) to use that knowledge to ask critical questions about the possibilities and limitations of making natural resource production more responsive to the concerns of a variety of publics. What is the nature of accountability in technoscientific organizations defined by divisions of labor? How do engineers embody and detach from the corporate forms employing them? How does the context of engineers' work position them to ask—or avoid asking—big, self-critical questions about the industries their work supports? What prompts engineers to step outside the limitations of their positionality, subjectivity, and expertise to seek out other forms of knowledge?[76] How can we chart more sustainable resource futures while acknowledging both the collective nature of the challenges we have inherited and our differential abilities and responsibilities to address them? How can that process be responsive to the distributed nature of industry insiders' work, without that acknowledgment becoming an excuse for the evacuation of responsibility? Finally, how can academics nurture the knowledge practices and professional ideals that would be necessary to open up more imaginative possibilities surrounding resource production, consumption, reuse, and waste?

METHODS

The book focuses on engineers who worked in the mining and oil and gas industries. Even within this relatively small professional network, there were noticeable differences in how they thought about and practiced social responsibility. Some of them created institutional change by integrating questions of social responsibility inside engineering decision making. A few dismissed terms such as *social license to operate* or *corporate social responsibility* as fads that should be managed by "social" people such as anthropologists like myself, leaving engineers to do their rightful work of technical problem solving and innovation.[77] The majority found themselves somewhere between those two extremes, recognizing the importance of public perception and identifying some ways that their engineering work articulated with improving it, without taking on broad institutional change as their own mission. While all these engineers expressed a range of opinions about their personal and professional obligations to address public concerns about their industries, they each had to account for their actions to multiple publics, from critics of their industries to friends, family, and the occasional anthropologist who was curious about their work and life.[78]

My research project—including the interviews, conversations, and fieldwork forming the basis of it—constituted one of the sites in which engineers practiced accountability to people outside their profession and their industries. In some cases my requests for interviews fit neatly within a company's efforts to expand its outreach and personalize the corporation by encouraging employees to open up and talk with people outside of the industry. This underscores that "ethnographic methods hold potential for plying into corporations' own self-representations."[79] Reflecting on how individuals and teams of personnel respond to research projects provides a window through which to study practices of corporate self-representation and accountability making. For example, the issue of company personnel restricting access to corporate spaces sheds light on how companies manage boundaries and self-representation, as "methodological obstacles" constitute "important knowledge about corporations."[80] In my own project, when my key contacts at a company suggested that I interview particular employees, that selection revealed much about what kinds of engineering and CSR practices they would like to circulate externally.

This project required rethinking traditional expectations for research methodologies, as "the collapsed roles of participant, observer, critic, employee, and colleague collide with one another."[81] During the focused period of research for this book (2014–2020), I engaged in activities that could be labeled *participant observation*, the hallmark of ethnographic research. I toured mines and well pads and accompanied engineers on their public engagement activities, such as the "science fair" community meetings hosted by oil and gas operators in Colorado (see chapter 4). I was able to participate in one corporate training event in which external affairs supervisors taught engineers how to respond to media requests that might damage their employer's reputation.

But *participant observation* does not capture how I have lived and breathed these topics since 2012 by virtue of my position as a professor at the Colorado School of Mines, an engineering and applied science university with long-standing and deep relationships with the mining and energy industries. Whereas the term *participant observation* conjures up activities that are distinct and cordoned off from one's "regular" academic or personal life, navigating an engineering and applied science university as an anthropologist made every day feel like a research day, requiring careful listening in order to understand my surroundings. When I participated in meetings, town halls, campus events, and alumni activities, it was with engineers and applied scientists, many of whom worked in the industries I was studying. When I attended the semiannual career fair, one of the school's most significant rites of passage, it meant talking with Mines graduates who were staffing the booths of more than 150 companies, most of whom had direct or indirect ties to the mining and oil and gas industries. When I taught courses, my students either aspired to work in those industries or were deliberately seeking work outside of them. I attended and organized campus lectures by engineers from industry as well as academia, including a lecture series specifically dedicated to CSR and engineering. I collaborated each semester with engineering professors across campus to integrate critical assessments of CSR into their own courses. I also led the social science research agenda for an industry-sponsored research center on water and unconventional energy, which provided opportunities to visit corporate headquarters and supervise research on public perceptions of fracking in Colorado.[82]

Yet, as Hugh Gusterson insightfully writes, "Participant observation is a research technique that does not travel well up the social structure."[83] Cultures of expertise inside corporations can be especially difficult to access, particularly when they are the target of external critique.[84] When regularized access was not possible in the study of Norwegian oil companies, for example, the research team developed methodological strategies characterized by "a high degree of personal flexibility, more semi-structured interviews than participation, non-continuous involvement with our interlocutors, mapping infrastructures of extensive geographical extent or opaque character, being present at or attending activities that involve alternative forms of socialites (social media, websites, documents, Skype-meetings, etc.), and even creatively designing situations [to] interact with and observe company representatives."[85]

My research also required creative research methodologies. I relied heavily on interviews, like other anthropologists studying relative elites, perhaps because "cultures of expertise are usually socially privileged, quasi-sovereign, often able to restrict ethnographic access, to monitor the acquisition and subsequent circulation of their expert knowledge, and even, if they are so inclined, to police ethnographic and theoretical content."[86] Nicole Smith and I formally interviewed around seventy-five engineers and those who worked with them, such as landmen (who negotiate lease agreements with mineral and surface owners), community relations practitioners, and lawyers.[87] The engineers came from different disciplinary backgrounds, spanning environmental, chemical, civil, and geological engineering, in addition to mining and petroleum engineering.[88] Though I did not formally interview any former students for this project, some of them did introduce me to their colleagues and supervisors. For example, I was able to interview about a dozen engineers and other personnel at one of Colorado's largest oil and gas operators after a student from my CSR course introduced me to a fellow alum he admired during an internship the prior summer. The Mines alumni network was valuable given the uniquely influential position the university plays in both the mining industry and the oil and gas industry.

Conferences became a generative research site, given that they are key "theaters of virtue"[89] in which executives and employees enact corporate forms while they generate and debate knowledge and "best practices" for a variety of activities, including CSR.[90] Conferences are spaces in which the people who work in these industries constitute themselves as a profession, sharing knowledge and nurturing professional and personal relationships over shared meals and drinks. I regularly attended conferences of the primary professional associations associated with mining and oil and gas activity. In addition to attending panels and social events, I actively participated by presenting my own research. This provided opportunities for receiving feedback from engineers and CSR practitioners as I was interpreting my data and crafting my conclusions, as well as for meeting more people to interview for the research itself.

As part of my larger collaborative work building CSR as a space of inquiry at Mines, I helped create an alumni interest group focused on social responsibility. For those who joined, the alumni group seemed to appeal to them as a new and meaningful way to connect back with their alma mater. Multiple generations of these alumni described their undergraduate years as being devoid of the kinds of social responsibility questions that interested them as students or came to occupy a place of great importance in their careers. They became visibly energized with the opportunity to participate in the life of the school once more. Some became participants in my research project, while others gave class lectures, judged student projects, and attended social events focused on social responsibility and engineering. These activities, coupled with my own teaching and mentoring at Mines, made my project one of "critical participation" in which I was thinking and working inside some of the social arenas I was studying.[91]

Even with these varied methodological tools and social relationships, I do not claim to present an all-encompassing treatise on engineers and accountability in corporate settings. I primarily interviewed people who had stayed in industry and made some peace with it, not those who chose to leave. I was also not able to "sit among" engineers as they did their work,[92] gleaning insights into their lives based on the small but significant details of how they arranged their desks and working days that are possible only through long-term, embedded fieldwork in offices.[93] Rather, my ethnographic practice revolved around deep listening to engineers and their colleagues as they thought through their work and lives with an interlocutor who was nonetheless at arm's length from them. What my interlocutors did share with me puts the things they did not share in sharp relief, creating a "negative space" that is present in its absence.[94]

One of the questions I was frequently asked by other anthropologists was how I could tell that my interlocutors were not simply feeding me a corporate line during our time together. How, they asked, could I tell that I was getting "real" data rather than a sanitized version of their thoughts and experiences suitable for public consumption? This line of reasoning makes assumptions about what good ethnographic data is. It privileges the desire to "reach behind the curtains" to access "backstage" interactions and to catch employees breaking rank to criticize corporate discourses.[95] The team studying Norwegian energy companies wisely observe that such confessionals are also a commonplace technique through which corporate forms admit failure and commit to doing better: "Even those moments of apparent spontaneous confessional, breaking ranks from the corporate line to admit failures of responsibility, impotence and frustration at the impending existential crisis of climate change for example, have become part of the ritual of public performance on the CSR/sustainability circuit."[96] Moreover, as I show in chapter 4, these assumptions about "good data" seem to hinge on notions of agency that privilege resistance rather than other ways of being in the world. Rather than interpreting my ethnographic data on a scale of authenticity according to how much an interlocutor was willing to criticize his or her employer or industry, I interpret all of these interactions as sites in which they were engaging in practices of accountability.

OUTLINE OF THE BOOK

The chapters that follow trace the practices of accountability through which engineers made sense of their work in the mining and oil and gas industries, to themselves and to their others.[97] While I sometimes group together engineers who worked in mining and petroleum, I also take care to separate them when their experiences differ. The different materialities of these industries matter for the "corporate social technologies" they develop.[98] For example, though petroleum engineer Aaron turned to mining cases to help him understand and manage the fracking controversies in Colorado, he also recognized the differences between those two industries for what a social license to operate might mean. Whereas mining was spatially intensive, oil and gas development was spatially extensive, sprawling over networks of roads and highways and interspersed with ranches and suburban developments. Whereas mines could operate for decades, the most intrusive periods of oil and gas development surrounded drilling and completions (the post-drilling process of making a well ready for production, which can include fracturing).

Chapter 2 provides an overview of the competing accountabilities that give rise to engineers' everyday practices. I suggest that the ethic of material provisioning and the social license to operate figure so strongly in how engineers understand their industries, their own work, and their collective responsibilities to the public because they seem to promise a reconciliation of the accountabilities engineers feel to their profession, to their corporate employers and clients, to the public, and to themselves.

Chapter 3 shows how engineers-turned-lawyers developed the concept of the social license to operate in order to shape questions about the accountability of natural resource production to be about *how* to mine responsibly, not whether to mine at all. By translating murky questions of public perception into questions about profitability, they were able to raise the stature and legitimacy of social and environmental concerns that had otherwise been peripheral to engineering practice and decision making. Yet as they opened up mining to greater public participation, they channeled it in ways that ultimately shored up the power of the company for which they worked.

Chapter 4 investigates the distributed agencies that characterize the corporate form, focusing on how engineers navigated their participation in an extended corporate "person." They did not encounter corporations solely as external entities that bore down on them. Rather, they moved between "enacting" corporate forms and distancing themselves from them.[99] This approach offers a new framework for thinking about engineers' agencies in the context of corporate work, going beyond the dominant ones that either condemn them for being conformists or celebrate them for being whistleblowers.

Chapter 5 theorizes the porous corporate form through the liminality of engineering consultants. I suggest that the professional autonomy desired by the consultants involved a narrowing of the publics, companies, and infrastructures for which they would be held accountable. While the language of choice used by the consultants to narrate their careers highlighted their own agency, they frequently found themselves hamstrung by their status as external "recommenders" for the projects. This liminal status ultimately served as a legitimizing device for the companies contracting them: consultants were widely perceived as being more "objective" and "independent" than the companies, even though in practice they remained financially dependent on those companies for their livelihoods.

Chapter 6 theorizes engineering pragmatism by analyzing engineers' practices of listening alongside their efforts to design industrial infrastructure that was responsive to public concerns. Faced with competing account-abilities, they tried to create industrial systems that minimized risk while providing financial gain for local people, as well as their employers. I argue that, while this practice of public accountability financially benefited some parts of the public while allowing companies to maintain or expand their reach, its focus on "actionable feedback" foreclosed broader questions about industrial development.

Chapter 7 concludes by returning to the questions of agency and accountability, reform and transformation, that weave through each of the earlier chapters. I make a case for developing new aspirations for engineers' accountability to help us collectively chart more sustainable resource futures. While the book proposes that we understand the accountability of technoscientific corporations through understanding the agencies of the people who constitute them, the epilogue explores how we might *alter* the accountability of technoscientific corporations by altering the agencies of the people who constitute them. There I also reflect on my own critical participation in engineering education, so readers curious about how my biography and institutional location influenced this research project should begin reading there.

2 COMPETING ACCOUNTABILITIES

Anthropologists have long studied accountability and account giving as a part of the "general fabric of human interchange."[1] While accountability is historically and culturally calibrated, taking different forms in everyday practice and institutional structures, Susanna Trnka and Catherine Trundle place issues of "responsiveness and answerability" at its heart. In their book *Competing Responsibilities: Moving Beyond Neoliberal Responsibilisation*, they show how multiple senses of responsibility, such as care for others and social contracts, exist alongside and sometimes in tension with dominant discourses steeped in more neoliberal notions of personhood emphasizing autonomy, entrepreneurialism, and risk taking.[2] While academics and my interlocutors often use the terms *responsibility* and *accountability* interchangeably—or define one in terms of the other—in this book I primarily employ *accountability*, as it more explicitly requires a specific person or group to whom one is answerable, whereas *responsibility* can be vaguer in not identifying particular persons or entities to whom one is responsible.[3]

Engineers' accountabilities to others are theorized and vigorously debated in the fields of engineering ethics and engineering studies. Engineering ethics is primarily but not exclusively the domain of philosophers, who use theory and prominent case studies of ethics failures to make normative claims about how engineers ought to act professionally. Engineering studies is more disciplinarily diverse, emerging from social science and humanities scholars seeking greater engagement with engineering than was occurring through the growth of science and technology studies.[4] Scholars working within and across these fields show that expectations for how engineers ought to practice accountability to their fellow colleagues and various publics are historically and culturally contingent.

This chapter provides a broad conceptual framework for understanding how accountability came to matter in the everyday lives of the engineers in my research. I argue for taking a practice-based approach that illuminates how engineers, in their daily life and professional decision making, navigate competing domains of accountability. The four domains I introduce here inform the rest of the book:

1. Formal accountabilities encoded in law, standards, professional ethics codes, and corporate policies
2. Accountabilities to professional ideals that, because of the history of the engineering profession in the United States, dovetail with institutional pressures to deliver and protect profit for corporate shareholders
3. Accountabilities to the publics that cohere around engineers' work
4. Accountabilities that are experienced as "personal" ethical frameworks but emerge from and reinforce broader histories and discourses, including industry

These domains often contradict one another, presenting engineers with vexing dilemmas. I argue that, in the absence of formal guidance for how to weigh, prioritize, or reconcile these domains, the engineers I came to know turned to two moral architectures that seductively promised to resolve the tensions among them: the ethic of material provisioning and the social license to operate.

EVERYDAY PRACTICES OF ACCOUNTABILITY

Dominant engineering ethics instruction in US undergraduate programs walks students through prominent engineering disasters, teaches them the ethics codes of relevant engineering professional associations, introduces them to ethical frameworks, and asks them to reason through which course of action they would take if faced with a dilemma themselves. While valuable, this style of teaching and learning addresses only a small fraction of the broad themes of ethics and accountability that engineers wrestle with in their professional lives. The engineers I came to know were called to account for their own decisions, as well as for their entire industries, on a daily basis. This calling to account happened when their family, friends, and acquaintances asked about their work in publicly maligned industries; when they wore clothing with company logos during off-work hours; when they encountered landowners or neighbors while visiting work sites; when they attended community meetings and public hearings; when they made presentations about their industries at schools or community organizations (figure 2.1); when they were approached by activists gathering signatures for legislation to more stringently regulate their industries; when they had to defend their decisions or proposals in team meetings; and when they were interviewed by me about their life and work. This account making also

Figure 2.1

An engineer who supervises fracking operations speaking to students during a high school chemistry class in Windsor, Colorado, in 2016. She demonstrated how the process works and argued to students that the technique, the accompanying chemicals, and the geological effects didn't harm the environment or public health. Students then got to handle the fracking fluid themselves. Photographer: Matthew Staver/Bloomberg via Getty Images, used by permission.

happened in more indirect and diffuse ways, such as when they read social media posts criticizing their industry, when they watched documentaries or news stories documenting their industry's flaws, and when they read summaries of grievance line calls or complaints lodged with the state oil and gas authority.

This book analyzes these *everyday practices of accountability*. By this term, I mean the practices in which the engineers became answerable to others for their actions, from their choices to pursue careers in controversial industries to their designs for the infrastructure to create natural resources, and they accepted, deflected, and reframed their own responsibility for matters of concern.[5]

This approach synthesizes an anthropological approach to accountability with the constructivist theory of social responsibility proposed by philosopher Deborah Johnson. She critiques dominant philosophical frameworks used to teach engineering ethics, arguing that they "decontextualize engineering and remain somewhat general in their implications for the social responsibilities of engineers" and instead advocating for viewing the "social responsibilities of engineers [as] constituted in social practices."[6] This approach has the advantage of compelling us to consider specific rather than generalizable actors, such as particular citizen groups rather than an amorphous "public." It also calls our attention to the specific expectations and understandings— shared and contested—through which explanation, justification, and judgment take place. Johnson identifies report issuing and whistleblowing as the two key arenas for engineers' practices of accountability, yet she identifies their weaknesses as failing to construct engineers as directly accountable for the health, safety, and welfare of the public. She argues: "Engineers are not required to explain or justify their behavior to publics until something goes wrong or until engineers—in the act of whistleblowing—bring something to the attention of a public. This helps to explain why so much attention has been given to the highly visible cases of engineering or technological failures. It is in these cases that accountability practices can be observed."[7]

My ethnographic research revealed myriad other moments in which engineers were asked to explain and justify their behavior to multiple publics, from backyard barbeques to requests made of the companies they represented. These everyday moments of account making may not be documented in public records, but they are crucial for understanding accountability as a set of social practices that shape and are shaped by dominant frameworks, such as international performance standards or professional norms. Analyzing the accountabilities of corporations through the everyday practices of the engineers and other professionals who enact them treats corporate social responsibility (CSR) as a dynamic and contested *field of relational practice*, not primarily as a static set of codes, guidelines, or standards.

In these everyday practices of accountability, my interlocutors attempted to reconcile contradictions among the four domains introduced above. They each had to justify their decisions based on the corporate policies of their employers, as well as the legal and regulatory standards governing their industries—the first domain.[8] These frameworks rarely provided clear-cut answers to their dilemmas but could be marshaled to support a position on a particular debate.[9] For example, Joe, a geological engineer who managed an exploration team in sub-Saharan Africa, made daily decisions that required him to calibrate company resources with desired community engagement outcomes, from choosing which economic development programs his team invested in to determining appropriate use of the company's trucks to transport villagers to the nearest urban center. Far from being black-and-white decisions, he saw how the established company practice of providing gifts and frequent transportation to the local chief, for example, could be interpreted as violating US anti-bribery law and professional ethics codes (the first domain), as well as his own personal sense of right and wrong (the fourth domain).[10]

Joe's professional ideals (the second domain) undergirded his conviction that technical professionals such as himself should use their knowledge to improve the lives of the less fortunate, chiefly by promoting sustainable local economic development rather than lining the pockets of elites. What drew him to geological engineering as a graduate student, in contrast with the geology and chemistry he studied as an undergraduate, was what he called its "applied" nature. Specifically, he wanted to use his professional skills to simultaneously create economic value for his company and the poor, rural communities who lived by their exploration projects. In recalling his time in Africa, he was most proud of the training he helped provide to local farmers to increase their yields, but felt frustrated that entrenched cronyism prevented his team from supporting broader economic transformations, such as by promoting the use of chicken manure rather than expensive imported fertilizer. It is through these daily decisions and activities—these *everyday practices of accountability*—on the part of engineers like Joe that a corporate form's CSR takes shape.

The primary dilemma is that, like most of the other engineers I met, Joe lacked clear guidance on how to arbitrate among competing claims to accountability within and among those domains. When he raised his critiques of the company's practices to his supervisors, he was told in no uncertain terms to drop the issue: "Don't worry about it," he recalled being told. Left on his own, Joe invested his efforts in activities and narratives that seemed to promise some reconciliation of those four domains. For example, he viewed the creation of economic value through farming and mining as not just the cornerstone of his employer's interests (first domain) and engineering professional ideals (second domain), but also as a service to the public welfare of the local community (third domain) and his own sense of right and wrong (fourth domain). "I still believe that the best way I can make a difference for [rural communities] is to actually find a mine to do my job," he said. "If I were to walk away and not find anything, it would be great for me to know that not only they remembered me fondly, but that also I had imparted some knowledge [about sanitation or farming] with them that would make their lives better." Whereas he found some peace with this vision of reconciled domains of accountability, positioning his company's interests and his professional ideals as indistinguishable from public interest marginalize other local calls for accountability or definitions of the public good. Indeed, Carl Mitcham provocatively argues that a "philosophical inadequacy" of engineering is that it holds an ideal of protecting public safety, health, and welfare without training its practitioners to know what public safety, health, and welfare actually are.[11]

INTERTWINED FORMAL ACCOUNTABILITIES
AND PROFESSIONAL IDEALS

Analyzing the four domains of accountability makes apparent the alignment between dominant formal accountabilities endorsed by corporate forms and the professional ideals for engineers. This alignment is a historically and culturally contingent achievement, not something "natural" to engineering itself. Formal accountabilities encompass regulatory standards, permits, corporate policies, negotiated agreements, international performance standards such as those endorsed by the World Bank and International Finance Corporation, and professional codes of ethics.[12] They are enforceable—though in different ways by different actors—and all have their own histories and exclusions that make them fundamentally political. Here I offer a brief overview of engineering codes of ethics because they are rarely considered in social science critiques of the mining and oil and gas industries yet provide a clear window from which to observe the coevolution of shifting professional norms for engineers and what scholars would identify as the economic interests of corporate forms.

The first engineers in Europe and the United States were military men, which engineering studies scholars argue embedded obedience to authority deep into the fabric of the profession. The link between engineering and the military was so prominent that the designation of "civil" engineers was precisely meant to signal nonmilitary applications of engineering practice. Historians and philosophers argue that, as engineers began moving from workshops and consulting firms into corporate employment around the turn of the nineteenth century, this ideal of obedience translated into loyalty to corporate employers.[13] In his fundamental *The Revolt of the Engineers: Social Responsibility and the American Engineering Profession*, Edwin T. Layton Jr. argues that the codes that first emerged during the Progressive Era ensconced the power of the pro-business engineers who advocated for loyalty to clients and employers at the expense of the more radical engineers who sought greater autonomy from corporations.[14] These codes of ethics were then translated into undergraduate engineering programs as early as the 1940s, when D. C. Jackson, an MIT engineering dean and consultant to electric utilities, chaired the Committee on Principles of Ethics of the Engineering Council for Professional Development.[15] Loyalty to employers and clients remained at the top of the codes of ethics until the crises of public confidence shook the profession in the 1980s, as described below.

Beyond loyalty to employers and clients, a second broad professional ideal of technocratic efficiency sought to find an internal good to engineering.[16] For now more than a century, engineers in the United States have prided themselves in their efficiency, perhaps most notoriously captured by mechanical engineer Frederick Winslow Taylor's attempts to scientifically manage the labor process by breaking apart workers' accumulated knowledge into small, measurable tasks to be monitored by engineers. This ideal of efficiency manifests in dominant expectations for US engineers to spend their careers in private industry and to dedicate themselves to ensuring what Gary Downey terms the "low cost, mass use" of their employers' products.[17] Unsatisfied by limiting themselves to the efficient management of resources, engineers have also sought to become managers for societal problems in general. Claims to social responsibility have been key to justifying this position: "Responsibility was the faculty of judgment that distinguished [the engineer] from the skilled laborer, the scientist, and the businessperson as he brought social progress through efficiency and invention."[18] The codes of ethics played a key role in publicly asserting and establishing internal ideals for engineers' social responsibilities.

A third and more recent professional ideal is the paramountcy principle: engineers should hold paramount the safety, health, and welfare of the public. It began to appear in codes of ethics as the professional associations attempted to address growing criticism of engineering in the wake of highly publicized technological accidents and disasters, including the Three Mile Island nuclear incident (1979), the Hyatt Regency walkway collapse (1981), the Bhopal toxic gas release (1984), the Chernobyl nuclear incident (1986), the *Challenger* explosion (1986), and the *Exxon Valdez* oil spill (1989).[19] The National Society of Professional Engineers, a bastion of professional autonomy among the larger pro-business engineering societies, was the first to bring the paramountcy principle to the top of its list in 1981, followed in 1990 by the IEEE (Institute of Electrical and Electronics Engineers), which in 2020 billed itself as "the world's largest technical professional organization dedicated to advancing technology for the benefit of humanity."[20]

Even as the paramountcy principle rose in importance, loyalty to corporate clients and employers remained. Matthew Wisnioski neatly summarizes this unresolved tension: "Lacking mechanisms of arbitration between employee and employer or even for upholding basic ethical codes, the rhetoric of responsibility in member societies elided conflict by portraying harmonious service to many masters at once."[21] In other words, the paramountcy principle acknowledges that engineers have responsibilities beyond their employers but does not provide guidance about how to reconcile competing demands between those public responsibilities and loyalty to their employers—a dilemma that continues to resonate in the everyday lives of engineers.

The tension between differing answers to the question of what engineering should be *for* remains unresolved nearly a century after the Progressive Era debates that seeded the first ethics codes. A persistent if underappreciated professional idealism has threaded through the history of the field.[22] In the 1960s and 1970s engineers critical of the military-industrial complex created professional spaces to link engineering directly with people seeking more sustainable local development.[23] In this period of social tumult, even "square" engineering scientists diversified their funding and research outside their traditional military and industrial sources to address the social problems of the day, from accessible technologies for people with disabilities to food poverty and renewable energy.[24] In the 1990s growing global attention to sustainable development began infusing curricular and extracurricular engineering programs,[25] and engineering educators made a crucial distinction between microethics that focused on individual conduct and macroethics that addressed systemic inequalities.[26]

These different answers to the question of what and who engineering ought to be for remain a vibrant field of debate and testimony to the multiple professional norms that exist in the field.[27] In the next sections, I analyze two dominant moral architectures that appeal to engineers because, I suggest, they seem to offer at least a partial reconciliation of those accountabilities to formal frameworks, to their employers, to their profession, to publics, and to their own senses of right and wrong: *the ethic of material provisioning* and the *social license to operate*.

ETHIC OF MATERIAL PROVISIONING

A steadfast and pervasive "ethic of material provisioning" animated how my interlocutors judged the ethicality of their work: they viewed their work and the industries it supported as valuable in providing the material foundations for consumers around the globe.[28] It is difficult to overemphasize the significance of this ethical framework, which pervaded interviews, conversations, corporate materials, conference presentations, and more. It positions engineers and others who work in industry as providing the "conditions of possibility"[29] for the everyday lives of people around the world. This sentiment was succinctly captured in booths at the 2019 Society for Mining, Metallurgy, and Exploration (SME) exposition that described mining as "making modern life possible" and made visible the minerals necessary for everyday technologies, such as smartphones (figure 2.2). This ethic stretches beyond

Figure 2.2
The ethic of material provisioning on display at the 2019 Society for Mining, Metallurgy, and Exploration convention.

any particular company. The National Mining Hall of Fame and Museum in Leadville, Colorado, for example, dedicates an entire exhibit to illustrating the minerals that make everyday life possible. This framework is an ethical one, as it imbues natural resource development—and the people who make it possible—with a sense of moral goodness. But like other ethical frameworks, it is inherently political. In this case, the ethic of material provisioning advances the mining and oil and gas industries' interests by justifying past, current, and future extractive activity.

This ethic of material provisioning was so pervasive that each of the engineers I came to know expressed it in some way, with some reproducing it in light of their own histories and others struggling against it. Here I focus on petroleum engineers. Laura described truly investing herself in her work when looking at the quality of life in "countries that don't have oil and gas" where "their hospitals, the power shuts on and off, and they run on a generator. [Here] we're able to have babies in incubators twenty-four hours a day and people on respiratory devices and all this stuff because we have a consistent power source. Solar can't provide that. Wind can't provide that. We have a better society because of oil and gas and coal. They are the moral, responsible thing." Laura's focus on sick newborns resonates with the discourses Hannah Appel critiques for positioning "hydrocarbon-based fossil fuels [as] the indispensable energy of social reproduction. . . . Without oil or coal—and by extension the people who produce them—birth itself, let alone transport, clothing, meals, electrical light, the entire American home, would be radically altered if not impossible as we know them."[30]

Whereas dominant criticism of the extractive industries lays blame on the people who work in those industries, the ethic of material provisioning distributes blame for the harms of mining and petroleum production throughout a more expansive network of people who consume products that come from minerals and petroleum. Invoking the ethic of material provisioning, then, seeks to create a worldwide "community of complicity"[31] that extends beyond those directly employed by industry. Kim, for example, believed that the negative effects of oil production should not be blamed solely on oil companies. She argued that those negative effects were "just our problem as humans, because we just wanted more petroleum."

She continued, "What I say to people is, 'Well, do you like to drive a car? Fly in an airplane? Drink milk? Coffee? Do you use Amazon?'"[32]

My interlocutors took great care to point out that even the people who criticized their industries depended on them. River, a mid-career petroleum engineer, had thought about the ethical dimensions of his work deeply, as it had generated intense criticism from his family based on their environmental commitments. He imagined himself having a conversation with environmentalists, saying: "Do you know the resources, the things that come out of the extraction process? It's your smart little socks. It's your Patagonia jacket. It's your skis. It's your contact lenses. . . . It's so integrated with society. It's not just the gas in your car, stuff like that. . . . It's not as evil as movies would portray." David, who had transitioned from engineering to executive roles, refused to buy clothing and outdoor gear from Patagonia because he viewed the company as taking a hypocritical stance in publicly opposing oil while depending on it (and its investors) for its business. And Abigail, an engineer at the very start of her career, complained about college friends who "drove their gas-guzzling SUVs to ski at Vail," a luxurious ski resort. Some were blunter in pointing out this dependence, such as a retired mining engineer who advised that critics of mining should be left to "live naked in the woods." By invoking the ethic of material provisioning, these engineers simultaneously cast positive moral valence on their own work—they were providing materials even to people who did not recognize or appreciate them—while expanding the "community of complicity" responsible for its harms to encompass the people who called the engineers' work into being by virtue of their consumptive practices.

As I analyze in chapters 5 and 7, some engineers who worked as consultants took a more critical stance on this widespread ethic of material provisioning, recognizing it as a defensive justification for their work. Yet even those who saw and critiqued the political narrowing of the framework continued to invoke it, using it as a justification for their own efforts to "steer the ship" and make mineral production as responsible as possible. It is telling that each of the professionals I met who sought to fundamentally rethink the ethic of material provisioning left industry entirely. In a sense, to work for industry—either as a full-time corporate employee or for a firm that contracted for corporations—was to understand one's accountability at least partially through this moral architecture.

Corporate Histories of Material Provisioning

While the engineers I met felt the ethic of material provisioning as a deeply personal ethical framework, it has a complicated history of its own. Companies, trade groups, and professional associations have long invoked an ethic of material provisioning to improve their public reputation. As early as the 1920s, the Anaconda Copper Mining Company adopted the slogan "From Mine to Consumer," and in the 1940s the company positioned copper as necessary for the suburbanization of the United States through advertisements such as, "There's copper in your radio . . . copper and copper alloys in your refrigerator, plumbing, and heating equipment."[33] By the at least the 1940s oil companies were engaging in similar public relations techniques, encouraging consumers to see how their everyday lives depended on petroleum products. After the boom in oil consumption by the US military in World War II, "the concerns of the industry shifted to the social construction of a 'postwar' American landscape that also was fundamentally dependent upon petroleum." During the 1950s, petroleum products "effectively saturated the whole of living, a set of practices linked to particular visions of freedom, domesticity, and health."[34]

Nearly a century since Anaconda's ad campaign, the ethic of material provisioning remains alive and well—and markedly similar. A competition at the 2018 SME conference to improve public perceptions of mining began by showing a promotional video from the corporate sponsor that extolled the necessity of copper for Americans' way of life, including "everything from wind turbines to diabetes test strips, cancer treatments, and car exhaust catalysts." The emcee invoked this ethic by saying, "Judges will be logging their scores on an iPad, which, coincidentally, comes from mining. Everything in an iPad comes from mining, right? So we're using the technology that comes from mining right here." While walking around the trade show at that same conference, I picked up bumper stickers declaring "If It Can't Be Grown It's Gotta Be Mined," "What's Yours Is Mined," and less subtle, "Ban Mining: Let the Bastards Freeze in the Dark." These were more humorous takes on the official posters, class lessons, museum exhibits, and other public outreach materials produced by the Minerals Education Coalition, the official education and outreach arm of the SME, that aimed to improve public perceptions of mining by making minerals more visible in everyday technologies and products.

The American Petroleum Institute also advanced the ethic of material provisioning by portraying oil and gas as providing "the molecular building blocks for products that Americans use throughout their day—from smartphones to fabrics to lifesaving pharmaceuticals. They're also essential to technologies and innovations that help solve some of society's greatest challenges."[35] Individual companies also invoked this ethical framework in their own advertisements. For example, ExxonMobil's "Enabling Everyday Progress" commercial showed the vast infrastructure and its attendant workers necessary to produce, package, and ship all the materials used by a middle-class woman boiling an egg in her kitchen on a natural-gas-fired stove.[36] The implication is that without ExxonMobil and its infrastructure Americans would not be able to engage in even the simplest of cooking activities.

The Politics of the Ethic of Material Provisioning

The strength and ubiquity of this linking of resource production and consumer lifestyles call out for analytic attention. Like other claims of technological inevitability, the ethic of material provisioning can be deployed to dismiss or deflect broader questions of social and environmental responsibility.[37] Anthropologists who encounter this reasoning in the oil industry in particular critique it as a defensive "ideology of inevitability" that forecloses broader questions of energy conservation and potential changes in infrastructure and consumer behavior.[38]

The ethic of material provisioning does justify industry activities by foregrounding the necessity of meeting current energy and material demands. And debates about the place of mining and petroleum production in our lives must include broader questions about the environmental and social costs of such activities, including the urgent issue of climate change. But analyzing how engineers themselves understand their work and the industries reveals another dimension to this ethical framework that should be integrated into debates about resource futures. My interlocutors drew on the ethic of material provisioning as part of their "deep story" about how they understand their place in the world.[39] It highlighted what they viewed as their noble contribution to the world they live in while critiquing their critics, specifically people who advocated for radical changes in energy and material production without understanding the material basis of their own privileged consumptive practices. While their senses of their profession and its place in the world were formed in the context of the wider messages disseminated by corporate employers and trade groups, their understanding of this ethical framework exceeded the official discourses.

The pervasiveness of the ethic of material provisioning may be attributable to how seamlessly it purports to reconcile the four domains of accountability. At the same time as it is felt and expressed as a deeply personal ethical framework that provides a positive moral valence to engineers' work, it also upholds dominant professional norms. First, the ethic of material provisioning expresses the long-standing industrial metric of "low cost, mass use" that shapes engineering in the United States.[40] While engineers who work in mining and petroleum production do not directly create-low cost, mass-produced consumer products, these outreach campaigns center precisely on making visible the minerals necessary for those consumer products, such as the smartphone at the SME tradeshow (figure 2.2). Second, the ethic of material provisioning casts a positive moral light on the corporations employing engineers: these corporate forms are in the business not simply of making profits but of providing energy and materials to people around the world. In effect, this framing proposes that the provision of energy and materials is a response to public demands and therefore is itself a public good. Where this interweaving of the four domains of accountability breaks down, of course, is when scholars and other publics assert different demands of corporations and the people who enact them: not to produce more materials but to slow or stop that production entirely.

SOCIAL LICENSE TO OPERATE

The ethic of material provisioning is experienced primarily as an ethical framework that provides a positive moral valence for engineers' work in the mining and oil and gas industries without—in the view of its proponents—offending the their accountabilities to formal standards, to their profession, and to the public. The concept of the social license to operate (SLO) similarly promises its adherents a reconciliation of those otherwise competing domains of accountability, but it is experienced more as a moral project more for industries and corporate entities rather than for the individuals who constitute them.

Social license to operate loosely refers to public acceptance, but the term is usually invoked without clear definition.[41] Advocates for the SLO define it as "the level of tolerance, acceptance, or approval of an organization's activities by the stakeholders with the greatest concern about the activity."[42] In the past two decades, the term *SLO* became "embedded within core mining industry vernacular,"[43] skyrocketed in academic attention,[44] and migrated to distant spheres such as synthetic biology.[45] A review of news articles for the term found that it appeared in fewer than ten articles a year from 1997 through 2002 but in more than two thousand articles in 2016 alone.[46] The seductive way in which the SLO offers to reconcile accountability to the public and the financial bottom line may help explain its quick update among industry actors.

Almost without exception, social scientists trace the genesis of the *SLO* phrase to 1997, when Jim Cooney, an executive at Placer Dome Inc. mining company, used it in a meeting with the World Bank.[47] Consultants specializing in managing community conflicts rapidly picked up the term,[48] which then became enshrined in the 2002 *Breaking New Ground: Mining, Minerals, and Sustainable Development* report commissioned by a group of the world's largest mining companies. That publication is widely considered a watershed moment in which executives publicly acknowledged that the industry's negative reputation was a problem and committed to sustainable development as a strategy to improve it. The term *SLO* appears six times in the report, though without definition.[49] The late 1990s also saw the first signs of SLO discourses in the oil industry. In the wake of the controversy surrounding Shell's plan to dispose of the Brent Spar oil storage buoy in the North Sea and outrage over the execution of Ogoni anti-oil activist Ken Saro-Wiwa in Nigeria, the company published its pathbreaking first

CSR report: "Profits and Principles—Does There Have to Be a Choice?" (1998).[50] The report is credited with popularizing the "triple bottom line" concept but also affirms a need to "balance between ensuring the commercial success of investments and our long-term responsibilities to society and the environment" before arguing that "it is essential to have endorsement from society—what some call a 'license to operate.'"[51]

By 2014 industry actors and academics publicized a business case for the SLO. The landmark "cost of conflict" study attempted to quantify how much money mining companies lost due to community strife, such as through shutdowns, delays, and opportunity costs for failed expansions. The staggering results—including the estimation that projects in operation phases lost $20 million per week of a shutdown—were published in an academic article in the *Proceedings of the National Academy of Sciences* and in an extended report published by the Corporate Social Responsibility Initiative at Harvard's Kennedy School.[52] Rachel Davis, a coauthor of the reports, was a fellow with that initiative after serving as a senior legal adviser for John Ruggie, the special representative of the UN secretary-general on business and human rights.[53] Coauthor Daniel Franks was a professor at the University of Queensland's Sustainable Minerals Institute, the other key institutional sponsor of the study. Their findings quickly became a ubiquitous point of reference inside industry, especially for people seeking to legitimize CSR teams, their activities, and their budgets. At one of the mining conferences I attended, the CEO of one of the world's largest consulting firms summarized the study for the audience by saying that it "translates environmental and social risks into real business costs and demonstrates that the loss of social license and ensuing cost of conflict really does have a financial implication."

By the time of my research, the SLO was invoked by almost all of my interlocutors. By emphasizing a high degree of community acceptance, the concept seems to address public accountability beyond what was formally required by laws and standards. Holding up a vision of mines and oil and gas installations that communities would actually welcome, instead of protest, spoke to many industry actors' personal ethical aspirations. And crucially, its basis in a business case seemed to suggest that engineers did not have to choose between their personal ethical aspirations and what was best for the corporate forms employing them: community acceptance was good for business.

The SLO and Its Discontents

There are compelling reasons why we ought to be skeptical of the SLO as a dominant image of accountability for these industries. Even academics and consultants who advocate for the utility of the SLO concept recognize its problems.[54] Whereas the concept draws its power from the formality and authority of legal licenses granted by governments, SLO proponents must go to great lengths to underscore the significant differences between them. The SLO, they argue, is not a tangible paper assigned through a standardized process but must be continually assessed and managed by gauging public perceptions. In the more sophisticated articulations of the SLO, public perceptions fall along a continuum of acceptance of the industrial project rather than being categorized as simple approval or disapproval.[55] The SLO is also vague when it comes to scale and to who ought to be granting it. While appeals to approval by "society" in general can marginalize the most vulnerable populations who bear the heaviest burdens of extractive activity, so too can locally focused stakeholder engagement efforts that ensconce the power of local elites.[56] The SLO requires companies to delineate—and, in the process, produce—a "community" that can deliver acceptance.[57]

Other social scientists offer more fundamental critiques of how the SLO gives the appearance of acknowledging broadened corporate responsibilities while ultimately bolstering the power of corporations. The SLO values the health of relationships with people—and sometimes people themselves—in terms of costs and benefits to companies and their investors, making it a form of private governance that entrenches market principles in domains that ought to be governed by rights-based frameworks.[58] It does not provide satisfying guidance on how to "navigate power inequalities, divergent interests, and diverse cultures of communication and governance."[59] In particular, it can skirt the special rights and processes accorded to indigenous peoples, as do frameworks such as free, prior, and informed consent.[60] Even John R. Owen and Deanna Kemp, known as constructive critics of mining, condemn the industry's use of the SLO on multiple counts. First, they argue that the SLO can provide a false sense of security by confusing the lack of observable dissatisfaction with approval. Second, and more damning, they argue that the SLO treats communities as a "risk" to manage in order to access land, which undermines the more meaningful stakeholder relationships necessary for sustainable development. They conclude, "Nothing short of a move away from social license at the project level is required to pave the way for a more proactive stance towards sustainable development."[61]

How the SLO Frames Engineers' Accountabilities

This section and the next build on critical scholarship on the SLO by illustrating an underappreciated dimension of the term's rapid rise: how it shaped how engineers learned about and attempted to manage their and their companies' accountabilities to people outside of their companies.

Mining engineer Austin said his first serious consideration of what he called the "social realm" took place during a "trial by fire" of controversy when he began working at a mine in Central America soon after graduating with his undergraduate degree in the early 2000s. The mine was under intense international scrutiny surrounding accusations that its personnel had abused the human rights of the indigenous people who lived closest to it. Austin arrived halfway through the initial construction, when tensions were already running high. He vividly described a confrontation that took place a few months before he arrived and colored his view of the project. Some major machinery was being shipped from the port to the mine but was held up when the caravan encountered a pedestrian overpass that was too low to allow it to pass. While the equipment was stranded a few hundred kilometers from the mine, he said, villagers started a "rumor" that it would be used to drain a large lake of significant spiritual and touristic significance.[62] He recalled that the villagers "took control of the truck" and were about to burn it when the police arrived. "People got shot. I mean, it was just a bad situation," he said. He described the incident as "really probably the biggest eye-opener I've had in my career as far as working in that kind of environment, where you really do need the social license to operate." He explained:

> I had never worked in an area with indigenous people. Adapting to their culture, to their values, was extremely important. I learned very quickly that the best way to relate with the people was, you go to their house. Even if you don't like it, you eat their food that they offer you. Drink their tea. A big piece of the social license is relating to the people, accepting their values, and something that was definitely not taught in school.

Many engineers wished that they had been offered more training as undergraduates on the thorny issue of community acceptance.[63] The SLO featured prominently in their descriptions of how they would explain the significance of the social dimensions of their work to novice engineers or students. Chris described himself as a "mining engineer at heart" despite having worked his way up to the executive level in health and safety, security, environment, and community relations. Though his first job in western Africa was a typical "technical" entry-level position for mining engineers, he soon found himself in charge of sociotechnical challenges he had not anticipated, including building roads and managing a large number of artisanal miners who were operating in his company's area. When asked how he learned about the "social" dimensions of his work, he candidly replied:

> I didn't learn. It was, go and do it. There were not books or anything else. It was almost a baptism of fire. The new general manager said, "Here comes the graduate mining guy. Let's give him the shitty job." I did give the blank look, but you didn't say no to your boss in those days. You just went there and you went, "Okay." So you drive there, you sit in the car, you think a bit, and you get out and you start talking. There was no preparation, no planning. It was all off the cuff, and wow, I made massive mistakes. One thing in community relations I can really tell you all the big mistakes, mismanaging expectations, bad relationships—all that I've gone through and done it. I didn't have any guidance in the beginning and made all the mistakes.

Chris was critical of the proliferation of terms to talk about public accountability. He learned, by transitioning from fieldwork to executive offices, that terms such as *CSR* or *sustainability reporting* may resonate with executives or investors but mean very little to communities. For him, *SLO* was one of the latest buzzwords. But when the conversation turned to teaching young industry professionals about the social dimensions of their work, the SLO and cost of conflict dominated his advice:

> If I was speaking to a group of young mining engineers or geologists, I'd say, "You know what guys? When I was studying what you were studying, what was more important was what was in the ground, the engineering. The world has changed rapidly. Now what we're finding is, you guys can find and engineer and have a feasibility study for the best project in the world. If you don't have that CSR social license in place, it's going to stay there. That's proven now, if you look at the research, most mining projects now that are stalled are stalled because of social reasons. We're talking megaprojects, multi-billion-dollar projects. You can throw all the money in the world at them, but they're stalled." My message to these guys is, "It's gone from being an afterthought, just like health and safety in the environment was years ago, to kind of being, it has to be front-loaded, guys.[64] It's not just getting it. You guys have got to live and breathe this. You've got to be globally aware, not just very good at the maps or the engineering stuff. If you can't do this, you're not going to rise up the ranks in the mining industry or be in a company that's going to be successful."

Both Austin and Chris used the SLO framework to interpret how they learned about their own and their companies' accountabilities to multiple publics. The engineers I met also used the SLO framework to try to convince others inside their companies to place a higher priority on their collective accountabilities to the most vulnerable people they affected. John, a mining engineer who ended up spending most of his career working on performance standard compliance in the oil and gas industry, expressed a common sentiment that the cost of conflict was instrumental for teaching reticent high-level engineering managers and CEOs to take community acceptance seriously. Describing the conflict that plagued the early days of a major infrastructure project in the South Pacific, he said that the managers refused to reconsider the company's dominant engineering practices until they calculated the financial cost of the delay:

> And when you're talking a day of delay in a gas project that ultimately will generate $12 million a day in gas production, it doesn't take long to do the

mathematics and even the village idiot can come up and say, "We would have been better off moving the gas plant." Had we simply moved the footprint to that plant, a couple of hundred meters to one side in the early stages, we could have avoided all of that. But the engineer said, "Wait, you can't do that. We've already made all of the drawings," and that sort of stuff. Now, in hindsight, they would have said, "Had we known what we know now, we would've done it differently."

The SLO was also the cornerstone of petroleum engineer Aaron's efforts to change how his colleagues thought about community engagement in the context of the fracking firestorm (figure 2.3; see chapter 1). He picked up the term from the CSR reports of international mining companies, but credited John Morrison's *The Social License: How to Keep Your Organization Legitimate* with helping him theorize the concept. Aaron found it useful to counter what he perceived as the oil industry's overreliance on the legal license (permits) and political license (political influence, e.g., through lobbying) to safeguard their ability to continue operating. By paying attention to public perception, he said, using the *we* of industry, "we started to see

Figure 2.3
Anti-fracking rally in Erie, Colorado, June 2012. Photo by Brett Rindt, courtesy Erie Rising under a Creative Commons license.

that we could make grand strides in improving the compatibility of the oil and gas development to preclude the friction or the problems." In speaking to my CSR course at the Colorado School of Mines, he described achieving the SLO as the key to addressing the "root cause" of the industry's public reputation problems. By growing the social license, he said, his team hypothesized they could stay one step ahead of battles over the legal and political licenses rather than facing them anew each election cycle.

Aaron, his team, and the other employees they trained found the SLO framework to be effective in persuading their colleagues to take the well-being of the people they affected, from the first stages of feasibility studies to operations. Marie described the SLO as "not something that we choose to do . . . but is part of the business now to get things done in the United States and then in the communities we're in." A petroleum engineer, she described herself as an "especially loud advocate of social license" and said she brought in the stakeholder engagement team to talk with her department because "a lot of people did not understand it," because they did not have experience working in places under intense scrutiny, such as Colorado. Marie argued for viewing safety, health, and social license as inseparable requirements of doing business. But she also underscored the financial dimension, linking the SLO with the "bottom line" and describing it as "a huge money driver for our company." Her comments point to how the framework of public accountability cast in the mold of SLO hinges on a vision of complementarity between doing the right things for communities and doing the right thing for a company's bottom line.

Harmony's Limits

The SLO provides a framework that helps industry actors make a variety of social concerns—from the most instrumental issues of public acceptance to more encompassing desires to promote long-lasting well-being—legible in corporate discourses that otherwise privilege financial accounting. In this way, the SLO seems to promise reconciliation of domains of accountability that could otherwise pull against each other: professional norms that position engineers as guarantors of profit for private industry as well as guardians of the safety, health, and welfare of the public; formal ethics codes, company policies, and international performance standards; public demands; and personal senses of right and wrong. But where engineers and others desire to see and promote compatibility, critics would point to the coercive nature of harmony ideologies and caution that some competing interests cannot be made complementary.[65] Indeed, the harmony among corporate and public interests proposed by the SLO can privilege the corporate ones, a process captured by the term *universalization*, in which "capital's interests come to subsume a range of issues raised by production and consumption."[66]

In reflecting on his career, Austin moved between recognizing the intrinsic ethical value of learning about and respecting people who lived very different lives and the business case for doing so. He did not view these two sources of value as being contradictory. The escalating conflict at the mine in Central America prompted him to try to better understand the indigenous people who lived around the mine, especially what he called their "values." While it was a personally meaningful experience for him to enter their homes, share food and drink, and learn about their lives, he also drew on the cost-of-conflict framework to justify those activities:

> A big part of it is that the companies have seen that, at the end of the day, the financial returns are what people are interested in. I mean, that's why the corporations exist. And they've come to realize that having a strong sustainability, community relations program actually pays dividends in the long term. You know, there's much less chance that you're going to have major issues, strikes, nationalization, things like that, if you have positive relationships with the people that work at your mine, that live around your mine, and the governments of the countries where you mine.

While Austin used the generic *people* to describe who was interested in financial returns, he specifically pointed to the importance of investors who would not invest in companies that did not have an SLO.[67] Using the *we* of the company, he said: "It comes down to money at the end of the day and what is our share price. And there's a lot of investors and funds now that won't invest in especially mining and oil and gas companies that can't really prove that they're being responsible. We've had representatives from some of these funds come to the mines to see if we're actually living up to what we say we're doing."

While Austin was willing to concede that some mining had been done irresponsibly, he argued that the industry's "evolution" had largely harmonized not just the interests of mines and the people who lived close to them but also his own ethical commitments with those of the company for whom he worked. When asked specifically if he ever disagreed with something his company did, he first justified its decisions to operate in socially and environmentally risky places and then described the (at least partial) alignment he felt.

> You obviously have to go where the resources are. You can't pick and choose where you want to build a mine. You can't pick and choose where the mineralization occurs. So in my experience, there hasn't really been a place we were mining that I feel wasn't a reasonable place to be mining. It's an industry where we do have an impact on the earth, and we can't always make it look the same when we're done . . . [but] I've been with this company [many] years now, and I guess I'm still with the company because their values are somewhat aligned with what my personal values are. I would say even more so today than when I started, because of this evolution that the mining corporations have had.

Toward the end of the interview, Austin took advantage of a pause in the conversation to note that his company had won various awards for its corporate sustainability reporting, suggesting that he seemed to view the interview as a chance to represent his company and perform his corporate identity (see chapter 4). Perhaps because of this, he was extra keen to portray the SLO as a harmonization of interests: by treating people and the environment well, the company could please its shareholders and maintain its own financial profitability. Randy, a senior mining engineer with significant international experience, made this harmonization clear when he offered a rare definition of the SLO in his interview: "We need to earn a social license to operate. So it means working with the communities and stakeholders that are impacted by our business or our operations, and earning their trust and proving ourselves as good neighbors, as a company that honors their heritage, their needs, but yet has a business operating."

Bev, a senior mining engineer turned executive who earned her stripes working on controversial projects around the world, also emphasized the importance of not just aligning interests but ensuring that potentially skeptical communities recognized the alignment of interests. She strongly believed that public perception was the industry's greatest risk:

> Threatening endangered species doesn't stop a mine. It's the public perception of whether it's acceptable to kill those threatened endangered species. It's people that stop projects. That's the only time you ever get stopped. If you are in deepest, darkest Africa and you're ready to move forward with your exploration drills and they surround your drills with people with sabers saying, "You're not going to do that here." There are way too many examples of those things. Call it social license. Call it what you want. But as far as I'm concerned, you've got to demonstrate to the local communities that it's in everybody's interest, what value you have to them over the long term. . . . You've got to demonstrate to local communities that you're of value to them long term.

While Bev subtly critiqued the *SLO* term—"call it what you want"—she invoked its key tenets of managing the risks from social conflict, caricaturized as African "people with sabers" stopping a drilling project. She universalized the interests of the company as serving the best interests of the community, saying that the company's success was in "everybody's interest." She believed that the legitimacy of mining companies rested on producing long-term economic value for local communities while minimizing environmental harms.

Value itself is a productively slippery term. While it can presumably encompass more than economic profitability, how mining companies operationalize the term often marginalizes social and environmental values.[68] As a case in point, the creation of economic value was also the primary "win" that engineers envisioned themselves as creating for affected communities in their efforts to quell the uproar surrounding fracking (see chapter 6). Aaron was pragmatic, saying, "You can either fight the social issues, or you can embrace them and try to solve them. But in the end, these companies are in it to make business and make money, and that's what they're there to do." His use of the term *they* instead of *we* subtly signaled his own distancing from the financial imperative he attributed to the corporate form and the people with the most power to direct it.

Decisions that pitted financial and other values highlight the cracks and fissures in the harmonious images promoted by the SLO and its proponents. Concern about the financial bottom line importantly constrained how engineers listened to and empathized with the people who criticized their industries. Petroleum engineer Emma vividly remembered a confrontation with landowners who threatened to not allow a rig onto their property unless the company paid them more money and made other concessions. The rig was necessary to plug and abandon the well, ending its productive life. The upset landowners requested to meet with the engineer in charge of the project rather than the company's landman (who negotiated the lease agreements with mineral and surface owners). Emma thought the landowners would be civil to her, given that she was visibly pregnant at the time, but felt personally attacked in her role representing the company as a whole. She recalled:

> It got very aggressive. I was very uncomfortable. When I told them I couldn't give them what they were asking for, they made a lot of direct, derogatory comments toward me. Then they tried to basically say, "Here are all the things we will take in exchange for letting the rig in." I told them I wasn't authorized to negotiate with them. And, "We don't cut deals like that and if you barricade the road, we will bring in the police. This is what will happen, because legally you can't do that. I can help educate you on where we are at and what we are doing, but I can't give you any money, I can't give you any pipe for your fence, I cannot give you gravel from the pad." They said to me, "Oh, you're one of them *smart* engineers."

The landowners insulted her professional expertise by sarcastically calling her "smart." Emma's use of the term *educate* suggests that she may have approached the landowner from a deficiency perspective, seeking to correct their knowledge or opinions of the engineering and legal dimensions of the decision that she and her colleagues had made. Whereas other engineers described direct engagement with people as stoking their desire to learn more about why they held the views they did, Emma's negative experience had the opposite effect. She remembered telling her supervisors, "I don't want to meet face to face with the landowners anymore. I think it's great, but I don't want to be in that situation again. It made me very uncomfortable."

Emma then provided institutional backing for her preference by highlighting the distributed agencies through which decisions were made in the company and communicated to members of the public (analyzed in detail in chapter 4). She emphasized that it was a "collective we, but typically it's an engineer's decision and we look at price." While she appreciated that the company had to maintain a good reputation in the community, she cautioned against going beyond their legal liabilities, as she worried that doing so was a financial burden and created heightened expectations that were difficult to standardize. To illustrate, she used the case of whether the company would extend power to its sites to run them off of electricity instead of generators. The benefit for landowners was a quieter site that had fewer emissions. After going to the county office and digging through files for every single permit approved in the past decade, Emma discovered that her company was the only one who proactively electrified each of its sites. This changed her and her team's decision making during the downturn in oil prices that had put more financial pressure on companies such as hers. She said:

> So when we come to a basis for a decision, if we are not required to electrify per the county regulations based on where we are located and where the power is, I've stopped proactively electrifying. Because it's the nice thing to do, but we don't have to do it. When we electrify a site we pay an additional $60,000 to $100,000 on a project to be the nice neighbor. So there's a lot more cost scrutiny now [with the downturn], and we can't afford to be that way—we won't be in business. So we are trying to fall back to what are the regulations. We need to be compliant, we need to meet them, but if it makes a landowner upset, I'm sorry, but this is what we are required to do, and this is what we are going to do.

The financial downturn, in Emma's view, provided justification for her philosophy that the company should focus on meeting their legal requirements rather than the specific requests of landowners that surpassed those regulations. In describing her experiences, she acknowledged limitations on the "win-win" solutions promoted by the engineers (see chapter 6) and underscored an image of engineers as protectors of their companies' financial bottom line.

Katharine, who worked full-time on a stakeholder engagement team, also admitted that it was difficult to balance competing interests from different audiences, even for an oil and gas company that had committed itself to the SLO:

> I think every business tries to reduce costs, the bottom line, but especially in our business, there's the short-term gains and the long-term gains. Yes, it may be cheaper now [to reduce drilling costs], but in the long run, are we even going to be able to operate? And that's where ballot initiatives and social license come in. Can we do enough now to mitigate our risks to actually show Colorado and show our communities that we will work with you? We have this duty to drill, you know, not only to our mineral owners but to our shareholders too. So we have to answer to a lot of people. Balancing that is hard sometimes, and sometimes there are things that we just cannot do. And making [those tough decisions] softer on the outside is our job.

These efforts at softening the impacts of industrial activity resonate with Marina Welker's description of CSR as one of the "ameliorative" disciplines that "cannot prevent the destruction of land, the pollution, and the waste production that are inherent in all mining. At best, they can mitigate these effects and help ensure that those affected are compensated. At worst, they facilitate more mining and the destruction that goes with it by overcoming resistance and unlocking 'high-value opportunities.'"[69] The "balancing" and "softening" that Katharine describes points to the contradictions obscured by the harmonious vision of shared economic, social, and environmental value promoted by the SLO.

CONCLUSION

Each of the engineers described in this chapter engaged in work that facilitated major infrastructural projects that came under strident criticism from others, from new mining and oil and gas projects in the Global South to firestorms surrounding fracking in Colorado. In everyday conversations, popular press, and academic scholarship, each of these projects has been attributed to named corporations, both large and small. While there are important rhetorical and legal reasons to attribute their ethical failures to "Corporation X" as a whole—for example, as many people refer to the "BP *Deepwater Horizon*" disaster of 2010 despite the vast archipelago of corporate forms implicated in it—this chapter argues for the importance of also disaggregating those corporate forms to see how "corporate accountability" emerges through the everyday actions of the people who constitute corporate forms.

When making decisions and making sense of their professional lives, the engineers I met grappled with multiple and sometimes competing accountabilities that emerged from the history of their profession itself. Dominant expectations for engineers in the United States hold that they work in private industry and provide economic value for their employers and clients. When a crisis of public confidence in engineers pushed members and leaders of the professional associations to include safeguarding public safety, health, and welfare in their ethics codes and expectations, they did so without providing guidance on how to reconcile this paramountcy principle with loyalty to their employers and clients. This meant that engineers continue to find themselves simultaneously accountable to corporate policies, to their profession, to the publics that cohered around their work, and ultimately to themselves. Oftentimes these accountabilities pull in different directions.

The engineers' desires to reconcile those competing domains of accountability help explain the ubiquity of two dominant moral architectures: the ethic of material provisioning and the social license to operate. Both aspire to reconcile what is good for a company's financial bottom line with what is good for the people affected by it. Like the field of CSR in general, they position companies as going above and beyond what is required by law for social and environmental performance. These moral architectures uphold dominant professional norms of engineers as protectors of corporate profit while expanding them slightly: in addition to their technical wizardry, engineers should nurture community acceptance to protect the financial bottom line. Yet even the ethic of material provisioning and the SLO do not fully reconcile competing domains or claims for accountability. Engineers and their coworkers could "soften" difficult decisions for the people they affected, to use Katharine's words, but they could not always meet both the demands of those publics and the expectations of their employers. The following chapters explore the variety of ways that engineers managed those irreconcilable differences, including detaching their sense of self and profession from the corporate form over the course of careers, projects, and conversations.

Social scientists would point out that both the ethic of material provisioning and the SLO may encourage engineers and others to speak *for* instead of *with* the publics that cohere around these projects. The ethic of material provisioning casts the public as in need of materials that engineers, through their work in private industry, can provide. The SLO ostensibly encourages deep stakeholder dialogue, but it ultimately aims at discussions of how such industrial activity can be done in a socially acceptable manner, not whether such activity should be happening in the first place. Chapter 3 shows that this framing of the social license as how, not if, industrial activity should occur was a deliberate goal of the engineers-turned-lawyers who in the 1960s and 1970s set out a business case for community acceptance, as they attempted to permit and plan controversial mining projects in the midst of the growing environmental movement and demands for greater public participation.

3 THE BIRTH OF A BUSINESS CASE FOR SOCIAL ACCEPTANCE

Social scientists mark the 1990s as the watershed moment in which the leaders of major mining and oil companies publicly committed their companies to being accountable to sustainable development and public approval, beyond the industry's already existing efforts in philanthropy, workplace safety, and environmental performance. In the popular retelling of this history, the key driver was the surge of conflicts mining companies faced while attempting to develop new mines in "greenfield" areas of the Global South without a history of heavy industry. The conflicts prompted the executives of mining multinationals to form the International Council on Mining and Metals, whose members pledged to uphold principles of "sustainable mining," such as ethical corporate governance, transparency, and human rights.[1] The group commissioned the landmark 2002 report *Breaking New Ground: Mining, Minerals, and Sustainable Development*, which called for companies to promote sustainable development in order to secure the *social license to operate*, a term that went undefined but generally referred to public approval of mining. The 1990s also saw an upwelling of anti-oil activism that took on Shell as a main target. Greenpeace rose to fame when its members occupied the Brent Spar oil storage buoy in 1995 to protest the company's plans to sink it in the North Sea. That same year the company also came under fire for its role in fostering political instability in Nigeria, vividly captured by the military's execution of Ogoni anti-oil activist Ken Saro-Wiwa.[2] In the wake of these events, Shell published its first corporate social responsibility (CSR) report in 1998, pledging to uphold a "triple bottom line" of people, profits, and planet.

While this history captures major mining and oil companies' wide-spread adoption of social-license-inflected CSR discourses, it overlooks the key period in which the concepts, practices, and institutional structures that would become the backbone of this field of practice were first developed: the 1960s and 1970s. This period was a boom time for the mining indus-try. In the United States, domestic uranium production peaked between 1960 and 1980. Coal production began a period of exponential growth as the 1970s oil embargoes sent the US federal government and energy companies scrambling to find domestic sources of energy.[3] AMAX Miner-als Inc., the diversified mining and metals multinational under analysis in this chapter, embarked on a $2 billion worldwide expansion beginning in 1969. This period of dramatic growth halted abruptly across the industry in the early 1980s, when commodity markets collapsed worldwide.

In this chapter, I examine archival materials and ethnographic inter-views to draw out one crucial thread of the AMAX history and its sig-nificance for broader, industry-wide transformations in how mining and then oil and gas companies manage their accountability with multiple publics.[4] I argue that as two AMAX engineers-turned-lawyers attempted to gain public acceptance for controversial mining projects in the 1960s and 1970s, they developed a concept of the business case for community acceptance; techniques to engage members of the public, from mine oppo-nents to nearby residents and politicians; and the institutional structures inside of the corporation to support their work. Though the AMAX actors did not use the term *social license to operate* themselves, the way they spoke about how they managed community relationships is all but identical to the post-1990s use of the term in industry. The role of AMAX and its engi-neers in crafting the business case for community acceptance, however, has been left out of dominant histories of the mining industry and its publics. Grappling with this history is crucial for multiple reasons.

First, the chapter shows how the now widespread practices of account-ability used to nurture the social license to operate were originally and explicitly predicated on a view of participation that facilitates rather than questions industrial development. Anthropologists and activists critique the industry's accountability practices for shoring up the power of corporations while attempting to disempower people and groups critical of their projects.[5] One of the techniques originally developed by the AMAX team—environmental impact assessment (EIA)—provides a particularly stark case in point. While scholars and activists have embraced the potential of the EIA process to "transform society" by encouraging dialogue and integrating environmental considerations into economic development projects,[6] it often it ends up preserving the status quo instead. For example, Fabiana Li argues that EIA prioritizes mining interests by "enabling corporations to define and ultimately enforce standards of performance," prompting those opposed to mining to go "outside" the document to voice their disapproval.[7] The archival and ethnographic material in this chapter shows that AMAX personnel originally and explicitly formatted EIA to facilitate the promotion and permitting of new mining projects, not to provide arenas from which to debate the value of mining and potentially halt it. This pragmatic approach to managing public participation is partially grounded in their engineering backgrounds.[8]

Second, the chapter positions engineers at the center of the mining industry's adoption of the social license approach to managing public accountability, a story otherwise told as one of executives in groups such as the International Council on Mining and Metals.[9] Stan Dempsey and Art Biddle, the AMAX personnel who led these efforts, originally worked as engineers before seeking out law degrees, and their efforts to make mining more accountable to its publics hinged on their embedding social acceptance in engineering decision making. Their engineering backgrounds mattered for multiple reasons. Like other mining companies, AMAX's executive ranks as well as mine site managers were dominated by engineers. Dempsey and Biddle could credibly speak the engineers' language while proposing social acceptance as a new constraint on those engineers' decision making. They tried to build bridges and translate knowledge among mining, law, and environmental conservation networks. Yet their engineering training and institutional locations deeply shaped how they imagined corporate accountability. They upheld a much longer naturalized relationship between engineering and for-profit business that positioned engineers as deliverers of corporate profit (see chapter 2).[10] Moreover, the public engagement techniques they developed invited greater public participation without fundamentally challenging either the power of the corporate form or the technical authority of engineers and applied scientists.

The chapter begins by providing an overview of AMAX during the period in question and Dempsey's pioneering 1960s "Experiment in Ecology" at the Henderson mine, which solidified his and the company's reputation for environmental stewardship and open planning. I then analyze Biddle and his team's work in the 1970s at the Minnamax and Mt. Emmons projects in Minnesota and Colorado, respectively, showing how they developed the concepts, practices, and institutional structures of a business case for social acceptance that would later become mainstreamed in industry as the social license to operate. Their work arguably made engineering decision making at AMAX more accountable to multiple publics by providing a more dialogic framework for engineers to engage their critics, but it simultaneously shored up AMAX's own power to frame debates about mining and provide industry-friendly solutions to potential problems, as Dempsey and his colleagues intended.

CALLS FOR ACCOUNTABILITY AND PARTICIPATION

The 1960s and 1970s US mining boom coincided with the rising tide of social movements that radically questioned the status quo. Science and technology studies (STS) scholar Christopher Kelty characterizes this period as one of increased demands for public participation:

> Activists, theorists, and politicians of the 1960s wanted a decentralized democracy of immediate, convivial, affective participation as a solution to the "organization man" and the anatomizing and alienating hugeness of midcentury corporate capitalism. Many observers in this period identified a problem in the technological systems and structures populated by an unelected technical elite, whose expertise, it was claimed, had come to substitute for the actual political debate central to a democracy. In the 1960s, such bureaucrats and technical experts represented a broken form of democracy—one that relied on expert judgment, couched in an inaccessible jargon authorized by the prestige of engineering, mathematics, and scientific management.[11]

These demands had particular implications for technical professionals. During the 1960s, people in the mining and oil and gas industries kept an eye on the growing environmental movement and its calls for greater public participation in decision making.[12] Stalwarts tried to keep business as usual, believing that they would be able to avoid government regulation. Andrew Hoffman describes this as a period of industrial environmentalism, in which "industry displayed an autonomous self-reliance based on technological self-confidence, and it viewed pollution as a problem it could handle itself. Government intervention was viewed as unnecessary, and environmentalists' concerns were viewed as exaggerated and not scientifically based."[13] Others, however, began experimenting with novel techniques for engaging with critics in order to stay one step ahead of changing regulatory requirements. In early 1960s California, members of an oil consortium predicted that they would not be able to access an oil field just off the coast of Long Beach without major efforts to camouflage its infrastructure, given the growing environmental movement and the still fresh memories of scarring oil development in Los Angeles.[14] They hired a Disneyland landscape designer to help design four oil "islands" that would extract oil in an aesthetically pleasing manner. They began producing oil in 1965 and remained active as of 2020, all the while appearing from shore to house modernistic colorful condos, waterfalls, palm trees, and shrubs.

Social unrest built to a fever pitch in the 1970s. The "decade of nightmares" witnessed a series of overlapping crises that called into question the place of science and engineering in society: the Vietnam War, economic "stagflation," political turbulence, urban strife, energy shortages, and environmental degradation.[15] The ample funding from the military-industrial complex that had built up science and engineering employment during the Cold War faltered during the drawdown in Vietnam and loss of public confidence. Countercultural "groovy" scientists connected their professional practice with the larger youth movement's interests in New Age spirituality and experimentation.[16] Even "square" scientists and engineers who were skeptical of if not hostile to the countercultural movement still adapted their research to accommodate the period's dramatic shifts in funding and public concern, from NASA's space-age meal delivery system for the elderly to the solar energy aspirations of Jack Kilby, a Texas Instruments engineer who coinvented the integrated circuit.[17]

To grapple with increased calls for accountability, some within industry continued experimenting with novel public engagement techniques. Leaders from the coal industry worked with the Sierra Club—the environmental organization that would go on to wage a powerful "Beyond Coal" campaign in the 2000s—to position coal as a "bridge fuel" that would phase out electricity from nuclear, oil, and gas sources.[18] The 1970s also witnessed major mining, oil, and construction companies taking part in a working group coordinated by Carroll Wilson, an MIT professor and founding member of the Club of Rome, which published *The Limits to Growth* in 1972, an enormously influential report that argued that resource depletion made sustained economic growth impossible.

The 1970s also ushered in a period of regulatory environmentalism, in which the newly formed US Environmental Protection Agency set down rules for industry and mediated between corporations and environmentalists.[19] The National Environmental Protection Act (NEPA) became law in 1970, raising the bar for the industry's environmental performance and public engagement techniques. NEPA mandated the EIA process to channel public participation in the assessment of proposed industrial projects, including mining, and by the 2000s EIA had become "the most widely emulated US policy in the regulatory playbook and preferred policy for managing natural resource use by connecting scientific evidence with public input in a in a highly regulated bureaucratic procedure."[20] Born out of the tumult of the 1970s, EIA is still used in the United States and around the world as a tool to involve the public in the evaluation of potential mining and other industrial projects.

EIA provides a "managerial approach to reconcile growth and environment."[21] The process collects studies, performed by third-party contractors, of existing environmental and social baseline conditions germane to the proposed project; uses those studies to predict a project's impacts; and then proposes techniques to manage the social and environmental risks identified in the analysis. The environmental impact statement (EIS) that emerges from this process is then opened to public comment, which must be addressed by the project's proponents in a formal response. EIAs form one trenchant site of "prognostic politics" that values some forms of knowledge over others.[22] While EIA has been hailed for its potentially transformative power to democratize decision making, the process and the documents it generates are critiqued by social scientists for ensconcing the power of the state and corporate actors who stand to benefit from a project's approval.[23] As summarized by STS scholar Javiera Barandiarán, "In practice, governments primarily use EIAs to improve industrial projects, not reject them."[24]

Within this broader historical context, employees and executives at AMAX developed practices of accountability that aimed to make engineering decision making more responsive to rising public concern without derailing the corporation's ability to continue operating and expanding. As the multinational mining company took on ambitious, controversial development projects in ecologically sensitive areas of the American West in the 1960s and 1970s, its personnel directly shaped what would later become EIA during their experiments engaging with mining critics and conducting baseline studies with ecologists who would go on to become prominent national policy makers. After EIA became a legal requirement, they also put together what they believe to be the first formal EIS for a US mine.[25]

The men who led the charge on these efforts were two lawyers who first worked as engineers: Stan Dempsey and Art Biddle. Neither fit the mold of industry executives, lawyers, or engineers. Both viewed listening as an ethical requirement for industrial development projects, and both spoke about the intrinsic value of nature, especially in a place with inspiring wilderness areas such as Colorado. But they also both firmly believed that industrial development could—and should—be done in ways that were respectful of ecologically sensitive places, and this belief infused the practices of accountability they developed and then promoted around the world. They accepted invitations to visit other companies' mines and corporate offices to advise them on the activities they developed that would later become commonplace in industry: conducting social and environmental baseline studies, setting up environmental departments and arranging favorable institutional structures to support their work, justifying environmental best practices to senior management, and managing increased demands for participation by government bodies, civic society groups, media, and critics. They opened their doors to journalists and academics to publicize their efforts. They presented their work at conferences around the world. They even shared their experiences Business and Industry Advisory Committee of the Organisation for Economic Co-operation and Development (OECD) in Paris. Dempsey would take the philosophy he honed at AMAX into a prominent career as a mining investor and executive.

AN "EXPERIMENT IN ECOLOGY" AT HENDERSON

AMAX was a major mining company on the global stage in the 1960s and 1970s.[26] Half of its overall revenues came from its Climax mine in Colorado, which then was the largest underground mine in the world, producing 75–85 percent of the global molybdenum supply. The ethic of material provisioning (chapter 2) saturated the mine's public outreach materials. As of 2020, the park that company personnel built across the highway from the entrance still displayed historic AMAX advertisements, which praise molybdenum's contribution to everything from lowering gasoline bills to making flame-resistant pajamas for smiling children.[27] AMAX was also a major player in the aluminum and coal markets, eventually becoming the third largest producer of each in the United States. The company boasted holdings in Africa, Australia, Southeast Asia, and the United Kingdom. Its corporate headquarters was located on the fifty-first floor of the towering Pan Am Building —now the MetLife Building—on Park Avenue in New York City.

Dempsey began his career as an industrial engineer at the Climax mine, where he conducted time-and-motion studies, optimization studies, and equipment purchase justification analysis. As an undergraduate he had trained for a year at the Colorado School of Mines but then transferred to the University of Colorado at Boulder to be closer to one of the small mines he already independently operated as an undergraduate geology student. After graduating in 1964 with a bachelor's degree, he went to work at Climax. Looking for ways up and out of the engineering job, he then obtained a law degree from the University of Colorado. He returned to the company as an attorney for Climax before quickly scaling the corporate ladder, thanks to his visionary approach for improving the environmental performance of mines through direct engagement with industry critics.

The key to Dempsey's early success was orchestrating the now famous "Experiment in Ecology" at the Henderson mine, located sixty miles northeast of Climax and fifty miles west of Denver. The ore deposit was located in an ecologically sensitive area of the Rocky Mountains, above timberline (ten thousand to eleven thousand feet above sea level) in the midst of alpine tundra that would be difficult to regenerate after the construction and operation phases. As the initial plans were being scoped, engineers estimated that the 0.5 percent ore grade meant that for each pound and a half of molybdenum they produced, it would require mining, crushing, and processing five hundred pounds of ore, generating massive amounts of tailings (leftover mined material) after the molybdenum was removed. That process was water and energy intensive, and the Henderson site was far from easy energy sources and located in a water-scarce environment.

Dempsey was an avid conservationist and active in the growing movement's social networks. Witnessing the rising power of the environmental movement, he drew on his personal connections to invite prominent ecologists to sit down with AMAX engineers and executives to collaboratively design the mine and its mill (material processing facility). His approach was highly controversial, as the majority of people inside of industry had placed their bets on ignoring the environmentalists or delegitimizing them. The main features of this unprecedented collaboration have been well documented by multiple sources, including AMAX itself.[28] Dempsey began making overtures to key players in 1964 to begin scoping out a potential collaboration. Under his leadership, the experiment was formally convened in 1967 by AMAX and the Colorado Open Space Coordinating Council, with participation from state and federal government agencies.[29]

The collaboration resulted in pioneering approaches to at least four areas of the mine design, some of which later became standard industry practice. The first was the location of the mill and tailings site, which are normally placed adjacent to the mine to reduce transportation costs. The site that suited both the AMAX personnel and the ecologists was located fifteen miles away, on the other side of the Continental Divide from the proposed mine site. AMAX agreed to build a 9.6-mile tunnel through the mountain to connect the mine and mill sites by rail.[30] The second was the routing of the power lines to bring electricity to the mine. The ecologists and AMAX engaged in first-of-their-kind conversations with the Public Service Company of Colorado to establish criteria for the design, location, and construction of the transmission lines. They routed the lines to avoid a rare stand of hundred-year-old Douglas firs and to pass through faster-growing lodgepole pine instead. The younger trees were taken out with horses, which were less destructive than trucks for the delicate alpine tundra, and the lines were placed by helicopters. Third, AMAX and the ecologists conducted what may likely be the first mining environmental baseline study—what Dempsey and others described as an "environmental inventory to establish quantitatively the nature of the territory that would be affected."[31] A few years later these would become the centerpiece of NEPA. Fourth, the land swap that made the mine and mill sites possible resulted in a net increase in accessible outdoors recreation area, opening up fifteen thousand acres to outdoor enthusiasts for nonmotorized activities, such as hiking, snowshoeing, and cross-country skiing.

The Experiment in Ecology had long-lasting influences, both for AMAX and for the industry as a whole. One was the mine itself, which opened in 1976.[32] An enthusiastic author in the *Journal of Metals* argued that the project's on-time completion alone qualified it as "an eminently successful engineering achievement."[33] Henderson went on to outpace Climax to become the world's largest producer of molybdenum, having produced 160 million tons of ore and 770 million pounds of molybdenum as of 2018. But an equally if not more important outcome of the experiment was the cultivation of an image of AMAX as an industry leader in environmental management. In 1969, as the experiment was under way, AMAX chair Ian MacGregor announced an ambitious $2 billion expansion campaign.[34] The cornerstone of their public outreach was what they called AMAX's commitment to developing resources in "complete harmony with sound environmental policies."[35] When making the case for more new development projects, AMAX officials repeatedly referenced the

Experiment in Ecology as evidence of the company's commitment to best-in-class environmental management and public engagement.

It was during this period that Dempsey began externally promoting the case that good public engagement was good for business. A 1974 speech captures this argument clearly:

> There is no question that AMAX is in business for profit. However, AMAX realizes that working within environmental constraints is profitable. AMAX's reputation as an environmentally concerned corporation has brought it business. For instance, a geothermal lease in Northern California was obtained by AMAX primarily because of its environmental reputation. AMAX realizes that sound environmental studies and interpretation with respect to exploration, development, operations and post-operations reduce expenses, delays and litigations and thereby increases profits and sustains its environmental reputation.[36]

That same year Dempsey wrote in a letter to AMAX CEO MacGregor that his travels around Europe demonstrated to him that "AMAX's environmental leadership can be a powerful sales tool for AMAX abroad."[37]

Dempsey's efforts to publicize the experiment and promote AMAX's activities were aided by Bettie Willard, a Stanford-trained alpine ecologist who participated in the Experiment in Ecology and went on to a long career in public policy. Willard was quoted describing the collaboration by saying, "AMAX officials knew that there were engineering limitations that could not be violated. And the environmentalists were firm in their convictions that there were ecological principles that should not be violated in the construction of a mine."[38] She became the director and then president of the Thorne Ecological Institute, where she founded a seminar series to translate ecology to policy makers. In a 1986 letter, she wrote that between 1967 and 1972 the institute hosted fifteen generals and fifty staff from the US Army Corps of Engineers and six hundred top officials from Congress, other government agencies, and private industry.[39] She then served as the first woman on the US Council for Environmental Quality (1972–1977), where she advised on the design of the Trans-Alaska Pipeline System. She then established the Department of Environmental Sciences and Engineering Ecology at the Colorado School of Mines, for which she received a UN

Outstanding Environmental Leadership Award. In that same 1986 letter, she wrote, "I think my greatest contribution has been to build bridges to non-ecologists, interpreting ecology and its utility to them."

In addition to the concept of the business case for community acceptance, Dempsey also made his first efforts to create novel techniques to directly engage would-be mine opponents in the process of mine planning. For the experiment, he invited scientists who had credibility in ecological circles but did not espouse blanket anti-industry sentiments.[40] This selection of participants had the crucial effect of framing the question of the Henderson mine as *how* a mine could be responsibly built, not whether the mine should be built at all. This effect was noted by an *Engineering and Mining Journal* article praising the collaboration between AMAX and the ecologists. Its author approvingly noted that the experiment meant that its participants "did not resort to legal structures, nor did they argue their case in the media or make emotionalism and panic the basis of action."[41] The author's bias is evident in describing critiques as "emotional," which not so subtly aims to dismiss and devalue those critiques. The author also privileged scientific and engineering knowledge by praising the collaboration for focusing on specific questions about particular environmental concerns—which tree stands would regenerate the most quickly, which tailings sites presented the least risk of water contamination, and so on—rather than broader questions about whether mining should happen at all.

Finally, Dempsey began creating new institutional structures inside of AMAX to support his work. He initially chaired an internal environmental planning and protection committee tasked with assessing and improving the social and environmental performance of the company as a whole, and he then expanded that committee into an entire corporate division that enjoyed institutional footing equal to the established ones dedicated to exploration or particular metals. Beginning in 1970 he served as director of environmental affairs and then added vice president to his title in 1977, when the division launched. His success in placing the environmental function at the upper echelons of corporate management occurred twenty years before this institutional structure became mainstream in industry.[42]

The division and its influence grew during the next two controversial mining projects: Minnamax in Minnesota and Mt. Emmons in Colorado.

OPEN COMMUNICATION AT MINNAMAX

As construction was under way at Henderson mine, AMAX personnel turned their attention to a copper-nickel mineralization in northern Minnesota. The region was home to existing taconite mines and to Babbitt, a small town on the eastern edge of the Mesabi Range named after the eponymous nineteenth-century mining investor. The region was also a cherished zone of environmental protectionism, home to a string of lakes at the Canadian border called the Boundary Waters Canoe Area. AMAX proposed the Minnamax development project at the same time as other major mining controversies were unfolding in the area.[43] Reserve Mining Company was under fire by environmentalists for dumping its waste rock into Lake Superior, and International Nickel Company was also proposing an open pit mine within a mile of the Boundary Waters Canoe Area. To manage external affairs at Minnamax, Dempsey hired Art Biddle, a young lawyer with engineering training and work experience.

Biddle graduated from the Colorado School of Mines in 1961 with a degree in metallurgical engineering, though he did not meet Dempsey there. Biddle first went to work for GE, including a stint at the Hanford plant in Washington, where he worked on manufacturing plutonium parts for atom bombs. After he was called to serve in the Army Ninth Corps of Engineers in Europe, he chose to attend law school at the University of Colorado instead of returning to GE, because he said he never could invest his "soul" in his engineering work there and wanted to return to his home state. Shortly after finishing his degree in 1968, a law school classmate who was already working at AMAX encouraged Biddle to apply to work there, suspecting that his engineering background would be useful. Biddle then met and interviewed with Dempsey, who promised him exciting work inside of the new Environmental Services division he was growing into a major force inside of the company.

Using the *us* of AMAX, Biddle recalled his first field assignment by saying that he "stepped off the plane in Minnesota and found twenty-six environmental groups opposing us." Shortly after he arrived, the state enacted a five-year moratorium on mining so that it could supervise a multimillion-dollar regional study of the environmental and social impacts of potential mining. The strategies he and his colleagues took to manage the public controversy were grounded in the same philosophy that Dempsey developed at Henderson mine. They made this connection explicit in their outreach materials and speeches to provide evidence for their appeals to public accountability. According to a brochure the company produced in 1980, AMAX sought to "refine a corporate policy on open communications" first established at Henderson. It went on to state:

> The effort [at Henderson] proved that there were many practical benefits from maintaining open dialogue with environmental groups, administrative and elected public officials at all levels, media, educators, industrial neighbors, and citizens interested in the social and environmental values affected by mineral development. The principles learned at Henderson were adopted as a doctrine and AMAX personnel in Minnesota have applied them in every possible instance. This openness includes sharing of mineralogical and environmental information, and ready access to the site for those who wish to observe evaluation activity and monitoring procedures. Hundreds of environmental leaders, reporters, public officials, teachers, and interested citizens have visited the project. Many more have attended meetings at which progress reports have been presented by the project manager of members of his staff.[44]

There were two key differences between Minnamax and Henderson. The first is that at Minnamax AMAX entered public debate before exploration began rather than after the characteristics of the underground mineralization were relatively well established. This meant that public engagement could not move immediately to collaborative mine design, as at Henderson, but instead focused on exploration, testing, and monitoring. The second was that by the mid-1970s company personnel could not simply convene an ad hoc committee with whom to experiment, collaborate, and negotiate. NEPA was in place, meaning that they had to comply with public hearings and subject their exploration project to potential scrutiny by a formal EIA. By preemptively going above the legal requirements for environmental baseline studies, AMAX was successful in its bid to avoid drafting an EIS.[45] AMAX personnel and the contractors they hired worked with the Minnesota Department of Natural Resources to design and implement first-of-their-kind baseline surface and ground water monitoring, given the significance of water contamination in the growing firestorm. They also collaboratively conducted some of the first studies on leaching from waste rock.[46] They cleverly turned these studies into external outreach opportunities, inviting state policy makers to the Minnamax site to witness firsthand the environmental monitoring taking place (figure 3.1).[47]

AMAX's environmental program won over all but the most vocal of their opposition, allowing them to sink a test shaft in 1977. The Sierra Club and Izaak Walton League, both conservation organizations, originally opposed and stopped AMAX from digging that test shaft but later withdrew their opposition after AMAX agreed to more frequent and extended water monitoring. In a press release the organizations commended the company for

Figure 3.1
Minnesota politicians and state officials seeing environmental monitoring in action while touring the Minnamax site, 1975. Photo courtesy Art Biddle.

The Birth of a Business Case for Social Acceptance

"good faith, complete openness and cooperation, and for the water quality monitoring already done."[48] The Minnesota Public Interest Research Group, the group most adamantly opposed to the potential mine, remained steadfast in their objections. AMAX personnel and the consultants they contracted were frank in the strategic choice of allies among the environmental organizations. As summarized by a civil engineer who did consulting work on groundwater and soil contamination, a part of AMAX's strategy was to "identify some environmental groups who were willing to listen and be reasonable" because AMAX personnel knew "they would never convince" the most oppositional groups. This strategy resonated with their strategic engagement of market-friendly ecologists at Henderson.

Biddle and his colleagues used the monitoring occurring at the Minnamax site to cultivate a reputation for AMAX as committed to environmental stewardship and community acceptance—a strategic move, given the variety of development projects they were pursuing during the aggressive expansion campaign. The front cover of a 1978 edition of the *AMAX Journal* (figure 3.2) featured a picture of water-quality monitoring at the Minnamax site, accompanied by the headline "Gaining Acceptance for Mining page 1." Photos of prominent Minnesota politicians and AMAX executives visiting the site for "Minnamax Day" illustrate the main story, titled "Mining and Community Acceptance." It summarized AMAX's strategy of gaining that acceptance through open communication about its environmental monitoring program. The article invokes the business case for community acceptance, which Dempsey continued using in his public appearances. At a 1978 meeting of the International Iron and Steel Institute, Dempsey argued that mining companies that develop skills and techniques in "environmental design and control, community growth management, and the handling of environmental conflicts ... will continue to prospect and prosper. Those who do not develop such capabilities will indeed have great difficulties in meeting their business objectives."[49]

In addition to the external public engagement techniques Biddle and Dempsey developed, they had to create productive ways to convince their fellow AMAX and consulting engineers working on the project to adopt—or at least echo—the external affairs team's philosophy of public engagement.

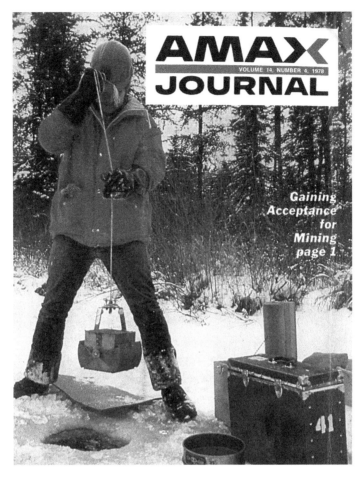

Figure 3.2
Front cover of the *AMAX Journal* from 1978. Photo courtesy Stan Dempsey archives.

Otherwise, departures from those values would call into question the external relations efforts premised on presenting the company as valuing open, respectful dialogue about mining, monitoring, and remediation. In other words, "external affairs" work required considerable *internal* work. Biddle vividly described the internal work as more emotionally draining than fighting mining adversaries in public: while he could understand the motivations of their external opponents and did not have close relationships with them, he said, "internally, you had to work with these guys."

Jack Malcolm, the project manager who was slated to become mine manager if the project went forward, had a reputation for being a stereotypical mining engineer. Biddle described him as initially being the "hard-nosed" type that would justify mining in traditional ways by saying, "We know how to get it done. We're going to increase your tax base by a whole lot. People are going have well-paying jobs. You'll like us in the end. Don't sweat the big picture too much." A former AMAX environmental scientist had a similar impression, describing Malcolm as an "old crusty mining engineer" who wanted to call all of the shots on the worksite. A consulting civil engineer who called AMAX progressive similarly described Malcolm by saying, "You call Central Casting, ask for a mine manager, and Jack would show up. Now, if you Googled a mine manager Jack's picture would be there. . . . He was a mining guy. After a few cordials he would say, 'You know, we're only doing this to get a mine built.'" These descriptions portray Malcolm a stereotypical engineer who was loyal to the company's bottom line and hostile to opening up discussion about the environmental impacts of mining to public scrutiny and discussion.

To encourage Malcolm to embrace the environmental and external relations programs as legitimate activities for a mining company rather than as threats to engineers' "rightful" ownership of technical decision making, Biddle encouraged Malcolm to accompany him as he attended formal as well as informal meetings with policy makers, local residents, and environmental groups. He hoped that listening to other people would help Malcolm see the bigger picture of mining beyond its economic dimensions. "I was responsible for working with him," Biddle recalled. "He was capable of talking to a government official, and they got to like him. But it took a couple of years before he really got with the program." With Biddle's coaching, Malcolm successfully hosted key people, including prominent senators and environmental activists, for tours of the test site, using the opportunity to learn about and address their specific concerns without being patronizing to them. "We spent a lot of time describing our external affairs project goals in our internal communications, measuring our progress in achieving those goals as carefully as we could, sometimes on a weekly basis," Biddle explained. "Over time this helped us educate technical project personnel and keep us on track." After the project ended, Biddle recalled that Malcolm apologized for giving him a hard time and thanked him for the work he had done for him personally and for the project as a whole.

The institutional structure Dempsey had built helped Biddle influence Malcolm to change the way he engaged with external stakeholders. Dempsey led the Environmental Services division, and Biddle reported directly to him, not to Malcolm. Malcolm worked for a different division and reported to a boss who was structurally equivalent to Dempsey. This placed Malcolm and Biddle on relatively equal footing in managing the project, with Biddle in charge of external affairs and Malcolm in charge of the exploration. Moreover, Dempsey had developed trust with senior AMAX executives by helping them navigate controversial projects, so he enjoyed considerable influence at corporate headquarters and could call on those executives to put pressure on his colleagues, such as Malcolm's boss.

Dempsey was a perceptive observer of corporate politics and found ways to work inside these to advance his division's agenda internally. For example, he and his group took advantage of the company's reporting structure to win over executives to their cause. "In reporting on our activities, we simultaneously taught them ecological principles so they would come to see the importance of our work," Dempsey explained to me, reflecting on his career. Rather than simply report that they had completed a study on estuaries for a site, for example, they wrote the report to teach its readers about the ecological significance of estuaries for the well-being of entire ecosystems. Thus, he and his team thus used strategic communication not just to understand and respond to their critics outside of the company but to advocate for their own projects and groups inside of the company and their operations sites. Dempsey was later invited to other mining companies and mine sites to advise them on how to set up environmental departments to handle the studies and permits that were increasingly required of industry. While these are now standard inside of companies, "no one had them at that time!" he exclaimed. "We had to invent them from scratch."[50] Dempsey also sent his personnel, upon request, to advise teams at other AMAX divisions and other companies, including their competitors. Indeed, AMAX's strategy of espousing corporate discourses of responsible mining to mitigate criticism of their projects would become part of the playbooks of the mining companies that proposed controversial new mines in the Boundary Waters forty years later.[51]

OPEN PLANNING AT MT. EMMONS

When Biddle left Minnesota in the late 1970s, he went directly back to Colorado to lead external affairs at AMAX's Mt. Emmons project. Geologists discovered the molybdenum deposit at Mt. Emmons in 1977, just after the test shaft was sunk at Minnamax. The Mt. Emmons project quickly became a matter of not just local but national debate centered on the nearby small town of Crested Butte, which was home to an eclectic mix of young, self-described hippies and an older population nostalgic for the town's coal mining days. Local politics were dominated by the younger people, many of whom had moved to the area to enjoy its plentiful opportunities for outdoor recreation, including at the recently opened Crested Butte Mountain Resort for skiing. W. Mitchell, the charismatic young mayor, galvanized a formidable opposition to the mine in Crested Butte.[52] Biddle recalled that Mitchell had "a radio announcer's voice that wouldn't quit." An airplane accident had left the former US Marine paraplegic, and over 60 percent of Mitchell's body was severely burned as a result of a motorcycle accident, making him a convincing David opposing the Goliath of a global mining company. Residents formed the High Country Citizens Alliance as a grassroots environmental advocacy organization to channel their opposition to AMAX.[53]

Mt. Emmons was located less than five miles from Crested Butte, and the subsidence—or gradual sinking of the land—caused by undergrounding mining would have been visible from the picturesque downtown business area. While the town of Crested Butte would have been the closest town to the mine, the county's socioeconomic consultants hypothesized that the county seat of Gunnison would experience the most population growth as a result of both the mine construction and operation. Gunnison County was home to twelve thousand people, with five thousand living in the town of Gunnison itself. The town was surrounded by ranching, and the county as a whole had a reputation for reliably voting Republican. Though Biddle and his team initially presumed that the "pro-business" sensibilities of the ranchers would make them easy allies, they conducted opinion polls and direct conversations with locals, which revealed deep concerns that the mine would set off uncontrolled boomtown growth. Affordable housing and good jobs in the county were scarce: the average housing unit cost $90,000, but the average annual income was just $13,000. This made the potential mine and its need to hire about two thousand people both appealing and worrisome for residents. After realizing that Gunnison was more welcoming to mining activity and that most of the social impacts of the Mt. Emmons mine would occur there, Biddle moved the AMAX office there from its original location in Crested Butte, where downtown businesses displayed anti-mining signs in their windows and the stained glass in the AMAX office window was punctured by a bullet hole (figure 3.3).

Figure 3.3
Display of the AMAX stained glass window punctured by a bullet hole, Crested Butte Mountain Heritage Museum.

As the controversy unfolded in Colorado, Dempsey ramped up his promotion of the business case for social acceptance, including inside of AMAX itself. In 1978, in the throes of both the Minnamax and Mt. Emmons debates, his team hosted an AMAX-wide conference on the importance of community acceptance and strategically located it at the Keystone ski resort to facilitate tours of the Mt. Emmons project. To bolster their credibility, the team commissioned a keynote address by Thomas Gladwin, then a business professor at New York University.[54] To an audience of AMAX executives, mine managers, engineers, community relations personnel, and lawyers, Gladwin argued that if AMAX could outperform its competitors in managing environmental conflicts, it would "reap substantial rewards . . . acquire access to secure, scarce and valuable natural resources in stable areas of the world that are denied to others. As such, constructive conflict management is a key to AMAX's profitable survival as a US-based natural resource corporation."[55]

AMAX executives adopted the business case in their own speeches. In his own address to the Keystone conference, AMAX president John Towers made a strong argument for the importance of social and environmental responsibility for a company's reputation:

> Public relations, public affairs and environmental affairs are key management functions and involve every member of management on every level. . . . A company's reputation determines its community acceptance and positively or negatively affects its ability to do business. Social responsibility—the concept that business must comply with the ground rules society sets and be a constructive force in society—is a fact of life. Our reputation rests on how we fulfill our social obligations. That reputation is like a three-legged stool. The legs are performance, communication and perception. Performance is delivering what we promise. Communication is telling the story of what we're delivering. Perception is how we're seen by others—our image and it has to be a good one. Take away any one of the three legs and the stool collapses.[56]

This more holistic way of viewing profit was also advocated by MacGregor, AMAX's president and CEO who presided over the period of rapid expansion and controversy. In an interview for the 1978 book *Footprints on the Planet: A Search for an Environmental Ethic,* MacGregor went on the record as stating:

> Carrying out a project in an environmentally correct way is a cost of doing business today. It's just good business to solve the environmental problems to the best of your ability before you start. Business is supposed to look at the quality of life for the people. My point of view is that we should look at the real bottom line—that is, we need that molybdenum in the 1980s and if we don't produce molybdenum in a way that is consistent with the environment, then we're not in business.[57]

MacGregor went on to clarify that he meant the "total environment, including the economic, the social, and the public understanding of what's good and what's bad." This articulation of interlinked economic, environmental, and social accountability anticipated the "triple bottom line" before the concept skyrocketed to popularity in the 1990s and 2000s.[58]

In addition to solidifying their internal company outreach, Dempsey's team formalized new techniques for channeling a surge of demands for public participation. In particular, they created a coordinated permitting process that also aimed to neutralize public opposition. The NEPA-mandated permitting process for the mine required between fifteen and twenty permits from separate agencies, each with its own requirements, applications, and public hearings, and each of which provided an opportunity for opponents to halt the project. Across the country, environmental groups had begun targeting these permits as opportunities to stop mines before they could be opened.[59] AMAX personnel were therefore eager to accept the Colorado Department of Natural Resources' invitation to serve as a guinea pig for a new streamlined permitting and public engagement process that would come to be known as the Colorado Joint Review Process (CJRP). Gunnison County and the key state and federal permitting agencies also participated.[60] Participants met once a month until the draft EIS was complete in the summer of 1982.

The CJRP was the most all-encompassing AMAX effort to create new techniques for public engagement (figure 3.4). It had two goals. First, it sought to streamline permitting by consolidating the social and environmental studies

Figure 3.4
A 1979 Colorado Joint Review Process meeting. AMAX personnel, including Art Biddle and Stan Dempsey, sit at the front left. Government agency officials sit at the front right. Photo courtesy Art Biddle.

and providing a single forum for public discussion. Second, it encouraged a more interactive format for evaluating industrial projects than was offered by the hearings mandated by the new NEPA regulations, in which members of the public registered comments on permits *after* they were submitted to the agencies with final studies and mine plans. CJRP meetings would invite discussion and input *while* AMAX officials and engineers were still in the study and design process, ostensibly providing an opportunity for the public to shape those studies and plans as they developed.

This kind of openness was unprecedented in the mining industry at the time. AMAX officials were willing to open up the planning process because they believed it would improve the design process while generating more public support for the project. Biddle's 1981 reflections on the CJRP summed up the public position he and his team took throughout the debate and that he maintained near the end of his life in 2018:

The generally accepted modes of public participation have proven to be inadequate because they involve the public only at stages when the major scope of a project can be changed but little, if at all. In addition, the means by which the public is normally involved—e.g., at public hearings, through the general regulatory review process, and in the courts—are adversary in nature and tend to encourage disagreement.[61]

The CJRP, in contrast, was designed to encourage what they called "open planning" (figure 3.5). Mike Rock, who directed community affairs for AMAX at Mt. Emmons, explained in a 1980 speech: "The CJRP provided an opportunity for the community to learn a lot and participate in actually designing the project. We have discussions about where the mill should go, or how the ore should be hauled . . . the CJRP provided the opportunity for people to participate in that discussion and influence our decisions."[62]

Figure 3.5
Close-up of AMAX representatives at the same 1979 public meeting, wearing shirts that say "Get High on Mountains—Crested Butte Colorado," which were presented to them from their adversaries in Crested Butte. Stan Dempsey is in glasses, second from the left, and Art Biddle seated next to him, third from left. Photo courtesy Art Biddle.

Looking back at the Mt. Emmons project in 2018, Dempsey argued that open planning meant that mining company personnel benefited by learning from others, saying, "If an ecologist has a great idea about conservation that is ultimately going to save me money and trouble down the road, I want to hear it!" This degree of public involvement in more open planning went a step beyond the open communications model at Minnamax, in which AMAX personnel, its consultants, and its collaborators at the state agencies made their collected data available to the public but did not involve the public directly in designing the monitoring projects or the potential future mine.

Presenting the public with engineering plans and financial analysis that were still in draft form prompted AMAX engineers and applied scientists to become adept at factoring social and environmental concerns into their analyses, decision making, and public communication. Some of these professionals became so skilled at this work that they took on public-facing roles to explain, answer questions about, and receive feedback on engineering plans. The primary debates centered on the mine size and the locations of the mill, tailings site, and ore hauling corridor.[63] While AMAX personnel had become experts in managing environmental controversies through the Henderson and Minnamax projects—and specifically brought in personnel who had worked on those projects to help the effort at Mt. Emmons—socioeconomic concerns about boomtown growth played a larger role at Mt. Emmons because AMAX's preferred mine plan would have doubled the population of Gunnison.

AMAX engineer Kay Ferrin took on the most public role in sharing how engineers inside of the company became accountable to broader social and environmental concerns. As project design engineer for AMAX's western operations, Ferrin participated actively in the March 1981 CJRP meeting (figure 3.6), which took place close to the July drafting of the EIS and therefore drew an unusually large crowd of about 150 people.[64] Biddle introduced Ferrin by describing him as a "sensitive engineer" and "professional ski instructor" who had worked for AMAX for twelve years in project design for new operations, including Henderson.[65]

Figure 3.6
Kay Ferrin presenting at the March 13, 1981, CJRP meeting. Photo courtesy Art Biddle.

Ferrin first addressed the issue of mine size, which had taken a central role in the unfolding debate because the type and size of the mine would determine the size of the workforce that would be moving to Gunnison County. The preferred alternative proposed by AMAX engineers, he explained, was for a mine that would produce twenty thousand tons of ore per day using the panel-caving method for extraction. In this scenario, the mine would last for approximately thirty years—but only if molybdenum prices remained steady. Peak employment would be fourteen hundred workers, including construction workers and production employees, but would add a total of about seven thousand people to the county when factoring in support services and families. To store the tailings from this size of mine, his team proposed that they be hauled fourteen miles to Alkali Creek using a railroad to be built by the company.

In explaining AMAX's preference for a large mine, Ferrin had to address a proposal for a "small mine" that was gaining traction. Some residents of Gunnison had formed a community group called Foresight to advocate for a mine with milder social and environmental impacts. One of its members was quoted in the local newspaper as saying, "We'll accept a mine, we just don't want one that is so big it will throw our community out of balance, so big that it will push out the ranchers, destroy the tourist trade, and endanger the feasibility of a college here."[66] The group contracted Alfred Petrick of the Colorado School of Mines to design a mine that would produce ten thousand tons of ore per day using the room-and-pillar method of extraction. This method would not create major subsidence, as would the panel caving method proposed by AMAX, and Foresight argued that the smaller mine size would require fewer workers. They had also stressed in the media that the small mine proposal would have a smaller area of impact and keep the mill closer to the mine, avoiding damaging the wetlands and wildlife at Alkali Creek.

To address the small mine proposal, Ferrin argued that while proponents of the small mine hoped that the slower production pace would extend the mine life, it would actually halve the mine life because the high-grade ore would need to be recovered first to recoup construction costs, and the room-and-pillar method would leave a greater percentage of the ore unmined. Turning to the issue of where to site the mine and mill, Ferrin began by explaining in plain language the process of mining and milling molybdenum and then went through the requirements and constraints on design: safety first, then reliability and efficiency viewed in a broad way to encompass "total life cycle costs," and finally the "responsible development of resources" or not "wasting [what] Mother Nature" had provided.[67] He then described how AMAX engineers and consultants went beyond simple profit motives in selecting their preferred mill and tailings sites: the tailings site must have a capacity of 300 million tons, but the engineers also considered each site's implications for flood control and downstream drainage, dam stability, visibility, and proximity to populated areas and ranches in narrowing a potential sixty sites to sixteen and then to just three. He explained:

Ideally, they would be right next to one another. Bing, bing, bing, you got a mine, you got a mill and just downstream of the mill you got a tailing pond, they are all in one short place. We looked at that. We could have built our mine right where [the ore] is . . . put a mill site just downstream from there and deposited all the tailing in the Coal Creek basin. That is possible from an engineering standpoint and it's really not too far out of line from an economic standpoint. Pure economics, the environment be damned. But that would have been an extremely irresponsible way for us to approach it because of the effect that it would have on the downstream communities, downstream on Coal Creek, downstream on the Gunnison River, and all of the other pre-existing uses of that basin. So that we threw out very quickly in the consideration of this ore body . . . then we have to look at the largest single impact area—the tailing deposition area. The reason that that's important is that that tailing deposition site is a permanent fixture. Long after the mine is closed down, the tailing deposition area will be there and we have to pick a site that is suitable for that kind of a long term, safe retention of the tailing that we produce from the mine.[68]

Ferrin then turned to ore haulage, integrating wildlife and other ecological considerations into his explanation of why his team preferred a railroad rather than a conveyor belt:

By going to the railroad instead of a conveyor belt, we are able to stay along the contours which puts us a little further down the flanks of the Whetstone and Red Mountain. And our wildlife studies have indicated that there is a significant amount of elk calving that takes place in this area up in here. And we would only—I believe it is in this area, [the environmental scientist] may be able to tell us exactly where it is—but in any case, this routing here skirts the edge of that along the edge of the elk calving area where our conveyor system would take us almost right directly through the middle of that elk calving area and summer elk range. So that brings us down to the edge of the elk range, and yet it keeps us above the human range. The humans have to range around here too. [laughter] They usually stay fairly close to the roads and we're above that and out of the view. Some of us, there are a few of us that wander out in the woods every once in a while. [laughter] But, this puts us above the roadway so the impact into most of the human environment is much less from up in here.[69]

Ferrin concluded by explaining the implications of different types of haulage systems for air quality and noise.

In the question-and-answer period that followed his presentation, Ferrin deftly answered questions from the audience. In response to a question on the accuracy of using construction workforce data from Henderson mine to predict needs at Mt. Emmons, for example, Ferrin first affirmed the value of the question and then explained the assumptions he and his team had used in projecting their numbers, drawing out differences in the two sites. Henderson's ore haulage tunnel was longer, he said, and more difficult to construct because it had to cross fault lines. He also said that the Mt. Emmons mine would be accessed by a horizontal tunnel rather than a vertical shaft as at Henderson, simplifying the construction process. These differences suggested that Mt. Emmons would require fewer construction workers. He then used the point to further call into question the small mine proposal, arguing that the number of construction workers for the small mine would not be substantially smaller than AMAX's proposal because it would require the same amount of facilities to be built.[70]

The Mt. Emmons case shows how social and environmental concerns rose in legitimacy and importance for engineering decision making. Ferrin stood out among the AMAX engineers for taking a prominent public role in the ongoing Mt. Emmons debate, eventually joining the community affairs team as an "invaluable addition in helping team members understand more about the construction process," as Biddle wrote in an internal memo.[71] But Ferrin was not alone. Many AMAX engineers and applied scientists attended the public meetings, and all those I was able to interview recalled that their engineering work was "special" because of the controversy. Reflecting on the project in 2017, Biddle said that even the technical staff at AMAX who otherwise would have taken an "old school" approach in dismissing the importance of community acceptance were aware of the unique nature of their work, given the heightened scrutiny the ongoing debate placed on each design, each study, and each proposal associated with the Mt. Emmons project. "What made these projects unique," he said, "is that everyone had to be able to justify *why* they were making each decision, each proposal, in light of public concern."[72]

In addition to bringing AMAX employees to public meetings, Biddle also cultivated employees to represent the project well in their social circles. This strategy, first encouraged by Dempsey at Henderson, would become commonplace in both the mining and oil and gas industries, as demonstrated in chapter 4. In a 1978 speech to a workshop on environmental management, Dempsey highlighted the importance of person-to-person outreach by employees by saying, "Every employee has a basic responsibility to help carry out our concern for the environment. Each one carries the AMAX message to their friends and neighbors daily."[73] In 2017, Biddle explained how he did this at Mt. Emmons: "We got local employees pretty involved. We wanted them out, you know, if there were community meetings they could go to them, and if they wanted to comment, they could comment. . . . This was a big deal that you read about in the paper almost every day." AMAX employees went so far as to create their own cartoons and public relations material not officially sanctioned by the company, including an eyebrow-raising bumper sticker that played on the double entendre of the company's largest mine: "What Crested Butte needs is a good Climax."

Neither the Mt. Emmons nor Minnamax mine was built. In 1982 AMAX received the permits needed to begin construction on Mt. Emmons, but the molybdenum market had collapsed, dropping from $25.50 per pound to $4.25 per pound in just two years. So did the markets for most other minerals, ushering in the long decade of stagnation and decline for mining in the 1980s. AMAX found itself in an untenable financial position: "AMAX, which had enjoyed sales increases of 43% in 1979 and a ninefold earnings increase over a ten-year period, found most of its businesses bottoming out. During the first nine months of 1981 profits were down 43%; the company's long-term debt was $1.2 billion; the following year would find it up to $1.7 billion."[74] AMAX announced in 1982 that it was "indefinitely postponing" the Mt. Emmons project. In dire financial straits, the company was merged with Cyprus and then was acquired first by Phelps Dodge and then by Freeport-McMoRan, which continued to operate the Climax and Henderson mines as of 2020.

CORPORATE FRAMINGS OF ACCOUNTABILITY:
HOW, NOT WHETHER, TO MINE

Through establishing the concept of the business case for community acceptance, techniques for public engagement, and the institutional structures and practices to support the work of community relations, Dempsey, Biddle, and their teams tried to make AMAX's projects and employees more accountable to some of the multiple publics who took an interest in them. Yet they did so all while shoring up the power of the company, as Dempsey, ever the shrewd businessman, did not hesitate to admit.

The primary way these practices of accountability bolstered the authority and reach of AMAX was by framing the question about mining to be one of how to mine responsibly rather than whether to mine at all. At Mt. Emmons this framing was a deliberate part of AMAX's overall external affairs strategy. In a 1981 memo to AMAX colleagues advising them on how to answer questions about the project's growing uncertainty, Biddle wrote, "The best thing we have going for us is the feeling that Mount Emmons is inevitable. We can increase this feeling of inevitability by our willingness to make expenditures in advance of the permit which demonstrates our commitment in addressing the secondary impacts of Mount Emmons and our confidence in our ability to obtain permits."[75] He continued:

> We may not expect overwhelming public support for our project, but we are trying to stress the inevitability of the Mount Emmons Project. This sense of inevitability can be increased by our efforts to show the community that we are fully committed to the Mount Emmons Project. . . . In addition we need to demonstrate to the community our willingness and commitment to address the secondary impacts of Mount Emmons. The less we demonstrate this and the more we require the community to move forward on blind faith, the more likely they are to impose strict mitigation requirements rather than to approach the secondary impacts on a cooperative basis.[76]

Public critique of AMAX's framing of how rather than if to proceed with Mt. Emmons escalated a year later when the EIS became public. Local government officials and residents identified a disconnect between the open communication and planning AMAX practiced during the CJRP process and the EIS, the legal document that emerged from the studies and meetings. The EIS mattered because it was ultimate document considered for permit and project approval. The work of drafting the EIS fell to Biddle and his boss, Ken Paulsen, a fellow engineer and Mines alum. They were given the herculean challenge of synthesizing the "more than 200 separate studies, 40,000 pages of study results, and more than 100 public meetings" generated by the CJRP.[77] They compiled a document that, even with seven hundred pages, did not include the full nuance of the studies and the debates they generated. They sent their draft document to the US Forest Service, the agency with the legal responsibility to draft the final EIS and submit it for public comment. The agency adopted all but one sentence verbatim from the one prepared by Biddle and Paulsen.[78]

The EIS received about five hundred pages of public comment, including thorough responses from government officials in both Gunnison County and Crested Butte. They criticized the EIS (and, by doing so, AMAX and the US Forest Service) for its narrow and strategic selection of information. Gunnison County officials expressed their appreciation for the open planning process but then asserted that the attempt to summarize forty thousand pages of study results resulted in a "mass of information . . . of little value to Board and Commission members who have followed and studied the earlier source documents."[79] They critiqued the simplification that occurred in the drafting of the EIS. For example, the EIS statement that the "project will bring about stable employment" misrepresented the actual findings of the socioeconomic studies, they wrote, and rested on an assumption of molybdenum prices staying high, even though the declining market had already resulted in decreased production and employee layoffs at Climax. Crested Butte officials were even blunter:

> As all of us are aware, a tremendous amount of information has been generated in order for the public and agency decisionmakers to fully understand the impacts of a major molybdenum mine. . . . Private and public money has been spent, a vast library of baseline data has been compiled, seemingly endless meetings have disclosed a broad range of concerns. It is with great disappointment therefore that we respond to a document that contains little analysis, that does not treat alternatives equally, that does not justify the choice of a preferred

alternative and indeed, reads like a thinly-veiled justification for AMAX's proposal. . . . This document neither helps public officials understand the environmental consequences nor does it 'foster excellent action' which protects, restores, or enhances the environment [as required by NEPA].[80]

NEPA mandated that an EIS consider not just the "preferred alternative" of the project proponent (in this case, AMAX) but others, including a "no action" one (in this case, for no mine to be built at all). Crested Butte officials argued that the EIS was written to make the case for AMAX's proposal rather than seriously considering the other alternatives, including their preferred one of no mine. Gunnison County officials agreed, likely thinking about the "small mine" proposal that was dismissed by the EIS as uneconomic but supported by many of their constituents: "Given the information presented in the Statement, the choice of one alternative over another is insupportable. There is no a priori basis for selection of the six alternatives; unfortunately, this means that only the AMAX alternative represents a viable and well conceptualized plan."[81]

The US Forest Service was legally required to respond to each of the "substantive" comments on the draft EIS but held the power to determine what counted as substantive and therefore to narrow the scope of what was debatable. A common theme of the public comments was that a mine was incompatible with the environmental values held by residents and that the existing regulations were not sufficient to guarantee environmental protections. In response, the agency wrote, "Debates about whether mining law should be updated to reflect environmental values should be resolved through legislative channels not through an EIS prepared under NEPA for a mining plan of operations."[82] Later in the document it stated, "Responses were made only to comments on the substance of the DEIS [draft environmental impact statement]. No attempt was made to address comments stating a position or an opinion, or to address non-substantive comments. This lack of response should not be viewed as a failure to consider a comment. These positions and opinions are valuable to the responsible officials, and will be considered in arriving at their decisions."[83]

This stance taken by the US Forest Service in 1982—that larger debates about mining and environmental laws did not have a place within EIS review—was one that Dempsey himself had long espoused publicly. The merging of stances on the scope of debate speaks to the influence AMAX wielded over the US Forest Service, which was already evident in the federal agency adopting the draft EIS by AMAX. Speaking in 1977 to the National Association of Manufacturers, Dempsey asserted:

> Rigid, mandatory impact assessment laws provide opportunities for outright opponents of economic development to attack project proposals in an inappropriate forum. Impact statements are developed to permit review of project compatibility with environmental quality goals and are not an appropriate vehicle to debate economic policy. Decisions concerning a nation's policy toward economic development, whether it relates to locating uranium fuel reprocessing plants or whether to permit construction of tourist resorts in the coastal zone, should be made at an appropriate level and should not be made in impact statements.[84]

In Dempsey's vision for public participation, such participation shaped but did not halt industrial development. In that same 1977 speech, he situated public participation in relation to financial value, a grounding that may have seemed natural to him as an businessman with engineering training:

> As I am sure you can tell, I am not an opponent of greater public involvement in environmental decision making. Quite the contrary, I believe that environmental activists have secured some results that are of value to business too. Opening up government has to be good for business in the long run. Making agencies more responsive will solve some of our problems too. At the same time, I feel strongly that we must work hard to make citizen participation positive and constructive, and that industry should vigorously oppose making such participation a vehicle for disruption and delay, a tool of cranks. . . . There are people who will do anything to stop a project. If their demand is legitimate, it will be vindicated by the legal process. If not, I can't see any reason to keep knocking ourselves out to communicate specially with someone who intends to use every communication against us.[85]

Dempsey concluded the speech by tying public participation back to the business case, stating, "I believe that in the long run such participation, properly handled, will be good for business."[86] He would make a similar argument about leaving major political decisions about values to Congress and the courts in a 1980 speech to the Business and Industry Advisory Committee of the OECD in Paris: "We believe that our projects are legitimate and contribute to the wellbeing of society and that if the public has all of the facts it will probably join us in that conclusion, and their judgment will be manifested in ultimate political approval of the project."[87] Paulsen, the engineer who wrote the Mt. Emmons EIS with Biddle, also attended that meeting and presented the specifics of AMAX's environmental monitoring activities, including the air quality and wildlife studies at Mt. Emmons. Whereas Dempsey tended to emphasize facts in his speeches, Paulsen noted the subjectivity of such studies and the impossibility of solving political debates through them, stating, "We frequently see an idealism which seems to believe that specific answers will result if only enough time, money, and effort are spent on study and analysis."[88]

While Dempsey valued a narrow scope of debate for providing a measure of certainty and efficiency in the midst of an increasingly tremulous regulatory framework and court of public opinion, some government officials and citizens who were active in the Mt. Emmons project likened the CJRP and EIS to what anthropologists would call an "anti-politics machine."[89] In 1982 a Colorado law firm that assessed the CJRP by interviewing its participants reached similar conclusions. They stated that the public remained an "unequal participant" in the CJRP and that the decision makers remained "more accessible" to industry than were the citizen groups. In the firm's view, the advantage of the CJRP for the company was that it "manages public participation in a controlled forum and encourages discussion of substantive, versus emotional issues." The law firm also saw "technical and substantive discussion of issues instead of just the emotional ones" to be an advantage to environmental and citizen groups who could then have a "larger voice in the process," though those groups might disagree with their characterization of value questions as "emotional." But crucially, the review also highlighted a significant disadvantage for those critical groups: "Because the CJRP is praised as the panacea for conflict resolution, continued community or environmental group opposition is less credible."[90]

Critiques of the CJRP foreshadowed what would become a prominent STS critique of participation, as eloquently illustrated by Gwen Ottinger's analysis of an oil refinery's "open dialogue" efforts that "actually opened up more space for industry authority."[91] Even as AMAX's public engagement efforts constituted a watershed moment for the mining industry by listening to the public and engaging its members in the design of industrial infrastructure, these processes ultimately advanced AMAX's own interests, precisely as their architects intended. The case thus represents an early and formative example of the much broader phenomenon of "participatory" dialogue processes reinscribing the power of industry. As summarized by legal scholars John M. Conley and Cynthia A. Williams in their Foucauldian critique of CSR conferences and documents, "To organize is to discipline, to control, and to limit." They continue by saying that what a company characterizes as the "value-neutral 'facilitation' of stakeholder dialogue can be seen as an exercise in control—control over who participates, how things get said, and consequently, if indirectly, what gets said."[92] AMAX's participatory processes did not silence criticism, as opponents maintained a public platform to register their disapproval, but they did shape the debate into one in which the key question was how, not whether, to mine responsibly, and the primary evidence considered in responding to that question was grounded in scientific and engineering knowledge and worldviews.

After Mt. Emmons, Dempsey went on to promote the gospel of the business case for community acceptance throughout the mining industry through site visits, conference presentations, speeches, and interviews. He helped disseminate his team's accomplishments in academia and business schools, such as through invited lectures and management case studies. He eventually retired from AMAX as vice president and also chairman of its Australian operations, directing substantial interests in Australia, Papua New Guinea, and Indonesia. Ever the entrepreneur, he went on to found the company Royal Gold through a reverse takeover and grew the company to a $1.5 billion market capitalization by the time he retired as executive chairman in 2008. As of 2020, Dempsey maintained an active interest in mining but also engaged in substantial philanthropy to support a variety of causes, including reforming health care systems.

CONCLUSION

Through the process of attempting to permit controversial mines and gain acceptance for them, Biddle, Dempsey, and their colleagues at AMAX assembled the key components of a business case for community acceptance. They channeled burgeoning calls for greater public participation into formats that shored up the authority and power of the corporate form itself: debates centered on how mining could responsibly proceed and were grounded in scientific and engineering knowledge. The legacies of their experiments in the 1960s and 1970s continue to resonate in contemporary practices of EIA and corporate public engagement efforts that emphasize stakeholder dialogue.

In this long and complex history, perhaps what is most crucial for the purposes of this book is the internal story: how Dempsey, Biddle, and their teams changed dominant expectations for engineers inside of AMAX to include responsiveness to public concerns in their engineering decision making. They mentored engineering personnel and operations managers to take humbler and more respectful approaches in listening to opponents, nearby residents, and public officials rather than simply rattling off the economic benefits of mining to them. They encouraged engineers and other employees to present a friendly, responsible face of their corporate employers to their friends and family and to share their work with audiences such as school kids. They invited engineers to contribute their knowledge and opinions to public meetings and hearings. And by taking the radical step of opening up mine planning to not just public comment but to public participation, Dempsey and Biddle fostered an ethos that engineering decisions had to be accountable to the concerns and expectations held by people outside the company, including citizens, organizations, and government bodies. Their success in modifying company expectations for engineers is likely because *they did not fundamentally challenge the dominant professional norm of engineers as creators of financial value for the corporate forms employing them.* Instead, they added social acceptance as a key variable in creating and protecting profits.

The social acceptance that they envisioned became standard for EIA around the world but reflected their particular positionalities as employees—and, in Dempsey's case, an executive—of a global mining corporation. For both Dempsey and Biddle, the hallmark of the public engagement practices they developed was what Dempsey described in 2018 as a "sincere commitment to listening to other people, including those we didn't necessarily agree with." He attributed the concept to Dale Carnegie's *How to Win Friends and Influence People*, one of the best-selling and most influential self-help books of all time, first published in 1936. It advises readers, among other things, to become genuinely interested in other people, be a good listener, and try honestly to see things from the other person's point of view. Sitting in his basement office, surrounded by a lifetime of memos, reports, and photographs, Dempsey smiled and said, "I'm coming to realize we used Dale Carnegie on the world." Indeed, in a 1977 speech, he said, "I think that that to successfully cope with the citizen participation demand, we must recognize some degree of legitimacy in that demand. Public relations techniques cannot mask hostility on our part. An insincere approach to citizen participation is self-defeating. . . . Dale Carnegie's advice goes for corporations, and the people who represent them. We need to put a smile on our corporate face, and take a sincere interest in people."[93]

Biddle also underscored the importance of respectfully listening to others, saying in a 2017 conversation with me that he always viewed the environmental groups opposing their projects "as a challenge, not the enemy." In a speech to other mining industry personnel during the early 1980s, when other mining leaders publicly insulted environmentalists, Biddle boldly stated: "If we want agencies and the public to hear about and understand the benefits of our projects, we must be willing to seriously listen to them and to involve them in the planning and programs which are designed to address their objectives and concerns."[94] Biddle recalled the friendships he cultivated in Crested Butte and Gunnison, including with people who positioned themselves in opposition to the project. He was remembered fondly by the people I met for his kindhearted approach to relationships. An engineer from one of the government agencies that monitored AMAX's activities in Minnesota described Biddle as a "down-to-earth kind of guy, flannel shirts in meetings, you know, you just instantly like him." Biddle's colleagues inside and outside of AMAX recalled receiving Christmas cards from him and his wife thirty years after the projects were halted and Biddle left the company to work as a construction attorney during the building of the Denver International Airport. Biddle later built on his experiences listening to divergent perspectives and trying to find common ground among them through volunteering for Peacemaker Ministries, a Christian group that used biblical principles to resolve personal and business conflicts, and serving as executive director for Conciliation Ministries of Colorado. Biddle passed away in 2018.

Along with their commitment to listening and sincere enjoyment of relationship building, Dempsey and Biddle each firmly believed that industrial development could—and should—to be done in ways that were environmentally sound. The pragmatic orientation that Dempsey and Biddle took to managing conflict is common among engineers but critiqued by social scientists for obscuring more radical questions from public debate.[95] Crucially, this worldview was directly embedded in the practices of accountability they developed to channel public participation in mine permitting. As this chapter has shown, these practices offered more opportunities for listening, engagement, and even collaboration at the same time as they framed public discussion around the question of how to responsibly mine rather than if mining should happen at all. The forms of public participation they developed ultimately benefited the corporation itself: their efforts established a more dialogic framework for engineers to engage their critics and other publics, but without significantly undermining their own authority in the processes of mine planning and permitting. These practices of accountability were originally and explicitly intended by their designers to facilitate mining activity and shore up AMAX's own power as a corporate form. This favoring of project proponents, such as corporations, in the EIA process has endured almost fifty years after they began their work, foiling what others had hoped would be a more fundamental transformation in how development is debated and planned.

Dempsey and Biddle were two technical professionals who had substantial but not unfettered power due to their institutional locations: Dempsey in particular rose to the executive ranks, and Biddle ran external affairs at both the Minnamax and Mt. Emmons projects. These positions allowed them more latitude to shape and transform the corporate form, especially its institutional structures, philosophies, and policies. But they had to work within an organizational setting, which meant that they could not just do as they pleased: they had to attune their actions to their peers and superiors while simultaneously trying to influence the actions of those others. Biddle's description of mentoring the curmudgeonly mining engineer to espouse a commitment to public dialogue is particularly vivid in this regard. Chapter 4 dives more deeply into this central quandary of organizational labor, but from the perspective of novice and midcareer engineers: to work as engineers, my interlocutors had to subject themselves to a division of labor that had direct implications for their accountabilities to their profession, to their employers, to the public, and to themselves.

4 CORPORATE SELVES

There can be little doubt that engineers derived substantial benefits from their alliance with business. But there was a danger that in gaining worldly things the engineering profession might have lost its own soul.
—EDWIN T. LAYTON JR., *THE REVOLT OF THE ENGINEERS*

Jay, a petroleum engineer in his midthirties, arranged the rock samples on his table while eying the crowd in the room. Two elderly white women wearing matching red lipstick and nail polish sipped lemonade and approached him. The company Jay worked for, a large oil and gas operator in Colorado, was hosting a community meeting in this wealthy gated residential development because they were seeking to develop new wells that would require hydraulic fracturing a few miles from some of the homes. The meeting was held in a clubhouse that otherwise provided a gathering place for senior citizens to play bridge, exercise, and attend other social events. Designed to quell concerns about fracking, the display behind Jay illustrated the depth of the wells by stacking seven images of the Empire State Building one on top of the other. It prominently labeled the "impervious rock layer" immediately above the well, the aquifer, and the cement and steel pipes used to encase the flow of natural gas to the surface. Wearing a company polo shirt, Jay greeted the women with a smile and asked them what they thought about the image. They asked if it was real. He assured them that it was an accurate portrayal of the subsurface and then invited them to touch some rock samples to feel for themselves how sturdy the layers were. They exchanged a few jokes about their contrasting levels of physical fitness, and then the women moved on to the next booth, saying to each other that people who thought fracking would contaminate groundwater were "simply uninformed." Jay smiled and scanned the crowd for who would approach his table next. "I love having a chance to get out here and talk with folks," he said to me. He had volunteered to do the event on his own time, he said, because he "believed in" what his company was doing.

Jay's booth was one of about ten set up in a U-shape around the room, each dedicated to a different concern, such as water, safety, pipelines, and leases. Each station was staffed by someone with expertise in that area—most often an engineer—dressed in an official yet casual polo shirt that seemed at home with the golf course visible through the windows of the clubhouse. Additional employees, dressed in identical company polo shirts, greeted newcomers at the door and offered to help them find the table of most interest to them. The employees easily outnumbered community attendees by a ratio of two to one. To engage the people passing by, each station included visual displays and hands-on activities, such as clear plastic cylinders filled with fracking fluid or flowback fluid, inviting people to pick them up, swirl them around, and examine them from multiple angles to get a more visceral sense for the fluids they had heard about in the news. Throughout the evening, people milled around from one station to another, sometimes stopping at a large open area with a generous buffet of appetizers and tall tables, intended to invite conversation. An official company public relations video extolling its social and environmental responsibility efforts looped continuously in the background.

On the heels of the 2010s fracking boom, this distributed "science fair" meeting became a key practice of accountability used by oil and gas companies in Colorado to manage their relationships with the people living closest to their operations (figure 4.1). The state was a central site in nationwide fracking debates. By 2020 it had witnessed a quadrupling of crude oil production in the previous decade and ranked in the top five states for oil and natural gas production.[1] Most of the country's then-major players—including Anadarko, BP, ConocoPhillips, Encana, and Noble—operated in Colorado along with hundreds of smaller firms. In 2018, the

Figure 4.1

A petroleum geologist answers questions from attendees of a community meeting in Colorado, July 2017. Photo by Paul Aiken/Boulder Daily Camera via Getty images.

Colorado Oil and Gas Association estimated that 89,000 people worked in oil and gas in the state. This oil and gas boom coincided with massive suburban growth. That same year, the number of Coloradans living in "petro-suburbs" was staggering: 429,000 people were estimated to live within one mile of an active oil or gas well.[2] The state rose to national notoriety for conflicts between the industry and activist groups seeking to halt it, and industry actors began experimenting with novel public engagement techniques, such as science-fair-style neighborhood meetings.

The power of these practices of accountability stems from the productive slippage between the corporate "person" and the human person representing it.[3] The community meetings positioned engineers to speak for the corporations employing them. To ask a question of the young polo-shirt-clad engineer was also to ask a question of the company whose logo adorned the shirt; Jay was positioned to speak and be recognized as speaking on behalf of the company—unless, of course, he marked his comments as departing from the larger corporate discourse with words or a wink. He and the other employees and contractors presenting that evening had gone through training to ensure that they stayed "on message," but they were also explicitly encouraged to share their own personal stories to facilitate more meaningful connections with the attendees. These meetings illustrated both the internally variegated nature of the corporate form and the work required to make it hang together.

This chapter investigates the distributed agencies that characterize the corporate form, focusing on how engineers navigated their participation in an extended corporate "person." Like other participants in collective life, engineers found themselves "not merely to be part of a collective, but actually to be that collective, or more precisely, to be an instance of a collective."[4] Acknowledging participation, to riff on Christopher Kelty's incisive critique of the concept, is not so much an ontological claim ("I am the corporation") as it is a moral or ethical one ("I am responsible for all of those things associated with the corporation, and the corporation is a symbol and instance of the things I am responsible for"). The vexing part of being "both individual and collective at the same time" is therefore the question of personhood and accountability.[5]

As employees, engineers found themselves held accountable not just for their own actions but for the actions of the multiple others constituting the corporate form. Those I met wavered between enthusiastically enacting corporate forms, folding their agencies together with multiple others and sharing in collective accountabilities, on the one hand, and attempting to detach their agencies and accountabilities from that broader entity when they found them to be compromised, on the other. This movement is vividly illustrated in the shift from engineers speaking with the *we* of the corporate form to referring to those entities as *they*.[6] Tracing how engineers managed distributed agencies and accountabilities and moved between enacting and detaching from corporate forms, I argue, provides a more nuanced understanding of engineers' agencies in the context of corporate work than those offered by dominant frameworks that either condemn them for being conformists or celebrate them for being whistleblowers.

AGENCY AND RESISTANCE

Agency has been a persistent and contested theme in social theory.[7] Anthropologists largely understand *agency* as the "socioculturally mediated capacity to act." This definition signals that these capacities vary by context and opens up space for the agencies of other-than-human entities and things, termed *actants* in actor-network theory.[8] Anthropologists resolutely hold that the agency of humans cannot be equated with free will, since agency is mutually constituted with social structure. Franz Boas, considered the father of American anthropology, articulated as much in his early observation that the "activities of an individual are determined to a great extent by his social environment, but in turn his own activities influence the society in which he lives, and may bring about modifications in its form."[9] In short, people's capacities to act are shaped by social structures, but social structures in turn are shaped by those actions. This basic insight animates what became known as practice theory[10] and poststructuralist theories of subjectivity.[11]

Feminist anthropologists caution against a common tendency to equate agency with resistance.[12] Saba Mahmood argues that "if the ability to effect change in the world and in oneself is historically and culturally specific (both in terms of what constitutes 'change' and the means by which it is effected), then the meaning and sense of agency cannot be fixed in advance, but must emerge through an analysis of the particular concepts that enable specific modes of being, responsibility, and effectivity."[13] In her ethnography of a women's Islamic piety movement in Egypt, the agencies Mahmood encountered sought not to disrupt the status quo but to embody virtues that Western feminists would associate with the reproduction of patriarchal systems of power. This led Mahmood to advocate for "delinking" the concept of agency from the goals of the progressive politics, arguing that such a link leads to the "incarceration of the notion of agency within the trope of resistance against oppressive and dominating operations of power."[14]

The question of agencies that do not resist is especially germane for studies of engineers and engineering. Edwin T. Layton Jr. and other historians mark the end of the Progressive Era in the United States, around the

1930s, as the watershed moment in which North American engineers relinquished professional autonomy in favor of stable, well-paying employment and management tracks inside of those corporate forms.[15] The context of corporate employment has been largely theorized as a tainting and stultifying imposition on the ethical agencies of their employees. In post-WWII United States, sociological and popular concern proliferated over the state of employees in increasingly large and powerful corporations and government bureaucracies, including foundational works in organizational theory such as C. Wright Mills's *White Collar* (1951) and William Whyte's *Organization Man* (1954). Engineers figure in this literature as icons of white collar conformity, loyal to their employers to a fault. Sociologist Gideon Kunda summarizes this view of the organization man as someone for whom "identification with the organization overrides all else and leads to the inversion of means and ends, a preference for conformity, a predilection for groupthink, a fear of creativity and initiative, and a dearth of ethics."[16]

This body of scholarship worries that people are not able to separate their sense of self from large organizations that seem not just to demand their time and energies but to "bind employees' hearts and minds to the corporate interest."[17] Or, as economist Richard C. Edwards wrote in his 1979 treatise on the transformation of work in the twentieth century, "Now the 'soulful' corporation demands the worker's soul."[18] Layton memorably expressed this fear for engineers in particular in the concluding sentence of his treatise on the compromised social responsibilities of the engineering profession, which serves as the epigraph for this chapter. This concern continues to be stoked by recurring revelations of practicing engineers who feel pressured to prioritize their employers' interests over the health, safety, and welfare of the public, as paradigmatically captured by events such as the *Challenger* space shuttle failure, the *Deepwater Horizon* offshore drilling rig disaster, and the Volkswagen diesel emissions scandal.

The persistent use of the word *soul* by Edwards, Layton, and others studying the plight of corporate employees is striking, and anthropologist Marcel Mauss's historical trajectory of personhood helps explain why.[19] Mauss traces Western notions of the "soul" back to early Christians, who added a metaphysical dimension to the Latin and Greek conception of the "person" as a "sense of being conscious, independent, autonomous, free and responsible" around the first century. He then argues that the concept of the "self" emerged in the sectarian movements of the seventeenth and eighteenth centuries as the experience of consciousness and reasoning. While the evolutionary aspects of Mauss's argument deserve the critiques directed at them,[20] this genealogy helps explain the tendency for academics and engineers themselves to use the concept of soul rather than others to critique corporate employment. In this lineage, the soul is the seat of ethics, the questions and judgments about expectations, accountabilities, and right and wrong.[21] And ethics is precisely what corporate employment, through its mandate for profit generation, is widely believed to compromise.

Scholarly work in the field of engineering ethics addresses the question of engineers' corporate employment by upholding the practice of whistleblowing, in which an engineer breaks ranks to call attention to unsafe or unethical practices. Yet embedded in this focus on whistleblowing are troubling assumptions about engineers' agency. As summarized by Deborah Johnson:

> Neither engineers nor publics seem to expect that engineers will blow the whistle or do something when they see a threat to the safety of the public. On the one hand, engineers who blow the whistle are often seen as heroes. On the other hand, those who stay silent are rarely held to account. It is as if publics expect engineers to be loyal to their employers when their employer's interests are in tension with the public good. . . . The ambivalence here is consistent with a long understood tension in the role of engineers; they typically work for employers and clients under the demands of a business environment while at the same time they have special expertise and the pressure to act on that knowledge. The latter is usually associated with acting as a professional.[22]

In engineering ethics' focus on whistleblowing, engineers are treated as agents of their own right when going against the corporate grain, with the implication that they are otherwise corporate automatons and therefore perhaps not even true professionals.[23] Yet as Gary Downey writes, "Agencies that do not challenge are still agencies, regardless of whether their performances of acceptance are discernibly active or passive."[24]

ENACTING THE CORPORATE PERSON

The phenomenon of engineers donning their corporate polos and putting their "selves"—their stories, their experiences, their expertise—at the service of improving the public perception of the corporate forms employing them is cut in the grooves of a much longer history of efforts to craft sympathetic "corporate souls." In the twentieth century these efforts frequently involved representing corporations in human form, ascribing the qualities of admirable human persons to the corporate persons they were intended to represent.[25] More recently, a team of researchers studying Norwegian energy companies point to the productive tension of the corporate person, stating that companies' power "relies on the embodied work of individuals who play between the scales of the personal responsibility and institutional responsibility" and are positioned to "embody the ethical agency of the corporation."[26] The engineers and applied scientists I came to know were all keenly aware of how they "played" these scales of responsibility as they engaged in practices of accountability, especially as they represented controversial companies and industries to a wide range of people, from their closest friends and family to strangers who publicly criticized their employers and academics who requested interviews with them. These practices positioned them as not simply employees *of* those companies but *as* the company, sharing accountability for its actions.

The engineers I met strategically managed their participation in an extended corporate person as they tried to set boundaries of accountability that resonated with their own ethical frameworks. At times in our interviews, they enacted a seamless integration of their personal mission and the mission of the corporate form. In part, this is likely an effect of the research context itself, as they may have viewed the interview as an opportunity (or perhaps requirement) to put a smile on the face of the corporate person—echoing Stan Dempsey's term from chapter 3—for the anthropologist asking questions to write a book for a potentially critical audience. A few interviewees were unwilling to admit facing any ethical dilemmas on the job, perhaps anticipating that even my best efforts to keep them anonymous could fail and expose their criticism of their employers or their own missteps. Others emphasized fissures in those larger corporate forms to disavow responsibility for the ethically unsavory behavior of their coworkers or managers. They seemed to relish diving into the gaps they experienced between what they thought was right and what "the company" eventually did. Most acknowledged that those ethical gaps existed but avoided going into detail about their specific content. Collectively, these materials underscore that interlocutors' interpretations of interviews and research projects shape what they are willing to share. But even the most "on message" interviews are ethnographically valuable, as they show the enactment of the corporate form in real time. To greater and lesser extents, engineers moved between enacting a corporate form and taking distance from it during a conversation, a project, or even a career.

These various enactments of the corporate person—as well as refusals to do this enacting—show that the senses of self held by my interlocutors were shaped but not determined by the corporate context of their work in industries that came under fire by others. This insight builds on Kunda's theory of organizational selves. In his ethnography of a high-tech US engineering company in the 1980s, he illustrates a range of tactics through which employees managed their relationships with their corporate employers. He argues that many engineers strove to create an "organizational self" by creating boundaries between their work and personal lives, by emotionally distancing themselves from the workplace, and by strategically embracing and distancing themselves from the organizational roles set out for them. But he also describes engineers who did invest their sense of self into the organization, embracing its norms and interests as their own. Importantly, rather than characterizing these employees as dupes who have swallowed a corporate line, he questions simplistic distinctions between "fake" and "authentic" selves to argue that senses of self are always crafted in contexts of power and control.[27]

One of the strongest ways engineers performed an alignment between their agencies and the corporate form employing them was to position themselves as personally advocating on behalf of their industries.[28] Kevin, for example, confessed that he knew that petroleum engineering was for him when he started dreaming petroleum engineering equations in his sleep as an undergraduate, including one instance in which he dreamed the exact problem that he encountered on an exam the following day. Young and ambitious, he wore trendy glasses that made him seem at home in downtown Denver's avant-garde coffee shops. He was married to an artist and enjoyed socializing with her politically progressive circle of friends. When we met, he had been working for about five years for an oil and gas operator known in Colorado for being on the forefront of community relations and environmental sustainability. He was one of the first volunteers to do formal and coordinated public outreach on behalf of his company, such as giving presentations in high schools about oil and gas. This same zeal infused in his social life, he explained:

> I'm pretty infamous for turning normal dinner double dates into hydraulic fracturing discussions. And usually involves taking the silverware and everything and using it as props. I can't stop talking about it most of the time, which is annoying for my wife, but . . . I've kind of chilled out. And once I've had the discussion with someone once or twice—once, probably—I chill out about it. Because I'm excited about what I do, it's easy to talk about it and get other people excited about it, too. I had a friend come up to me or tell me the other day that she ran into a petition gatherer for an anti-fracking initiative. And she was like, "I told them I think I'm for fracking." And I was like, "There we go. I'm glad we knew each other, because, you know, hydraulic fracturing is such an obscure topic for most people." I, at some point, had a conversation with her, and she was willing to defend me and the industry.

In his narrative, Kevin placed his own outgoing advocacy at the service of the corporate form employing him and the industry at large, linking their accountabilities. He was like other young engineers I met who were less than ten years out of their undergraduate degrees and enthusiastically "played the scales" of personal and corporate accountability. To challenge people's preconceptions of their industry, they made their position inside of the oil and gas industry—and their status as enactors of their corporate employers—hypervisible in social circles that might otherwise be hostile to them.

Abigail, similar in age to Kevin, described the same propensity to represent her company and industry among its critics, especially from her own social circles. Describing herself as a petroleum engineer, feminist, and outdoorswoman, she said that she chose a career in oil and gas after attending a left-leaning liberal arts college because she was compelled by the "disconnect there between people who worked in the industry and the public." Her engineering friends, she said, "opened [her] eyes" to the oil and gas industry while her liberal arts peers demonized it "in the same breath as they drove their SUVs up to their condos in Vail to ski." After graduating with a master's degree in petroleum engineering, she chose to work for a company she viewed as "progressive" so she could "be a force from the inside" and make immediate improvements for the people and environments affected by oil and gas development. After explaining how she designed well pads with smaller footprints and quieter machinery to be more respectful of the people who lived closest to them, she said, "We need people working on renewables, we need people doing research, but ultimately their solution is not going to make an impact probably in our lifetime. . . . But the impact that you can do today, a petroleum engineer or an engineer in oil and gas, you can see the impact of it tomorrow. So that's super powerful for me in terms of social responsibility and sustainability."

Abigail poured her time and energy into volunteer opportunities, proudly wearing her company polo shirt to events as such as those encouraging girls to consider STEM (science, technology, engineering, and math) fields. She was one of the most active ambassadors of the Society for Petroleum Engineers I met, serving on multiple committees and supporting their outreach efforts on her own social media accounts. In their interviews, Kevin and Abigail portrayed their public enactment of the corporate forms employing them as simultaneously an expression of their own self. This intentional intermingling played the scales of responsibility, attributing their own ethical commitments to the corporate form and even industry as a whole.

Intermingling Environmental Ethics

Throughout my research, this intermingling of the engineers' selves with the corporate forms employing them often seemed to be done with an eye to bestowing an engineer's environmental commitment to the corporate form. Chapter 3 describes how Stan Dempsey and Art Biddle engaged in this practice when they positioned themselves and their employees as outdoor enthusiasts; if the head design engineer, Kay Ferrin, was a ski instructor, for example, the implication was that he would not propose mine plans that would endanger the ecologically sensitive alpine habitat.

This attribution of environmental ethics to corporate forms via the people enacting them was ubiquitous by the time of my research. Dan, a geologist, said he welcomed work for a mining company because it provided him a stronger platform to enact his environmental ideals. He described himself as becoming fascinated by mine reclamation early in life, as he went hiking in the mountains, finding abandoned mines and seeing the stream pollution they caused. His initial work for consulting firms on surface mine reclamation was frustrating, he said, because clients pushed for the minimum necessary, which offended his own senses of public, professional, and personal accountability. "Do I want to put my name on something?" he asked himself. "Names are what goes in the newspaper. Names are what goes on your reputation."[29] Whereas he felt a strong desire to separate his sense of self from the work that he did as a consultant, he said that as a corporate-employed environmental scientist he felt empowered to put his own dreams into action and be held accountable for them. For example, he supervised the planting of wildflowers and aspens on reclaimed mine property alongside a highway frequented by outdoor enthusiasts. "Every time I see a photographer taking that shot," he said, "I feel a little proud about it."

Dan's experiences resonated with multiple ranchers and outdoorspeople in my previous research who described becoming environmental scientists and engineers in Wyoming coal mines so that they could participate in reclamation efforts. One explained that she got the "warm fuzzies by going home with the knowledge that the critters, the natural grasses and everything else are going back like it should be." A public relations campaign by the company she worked for played up the environmental commitments of their employees as evidence of the company's approach to environmental responsibility as a whole. In one ad, a fellow engineer explained that she could be an "environmentalist" who is proud of her family's ranching livelihood while working for a coal company because the company was committed to doing the right thing. These statements again play the scales of accountability, aiming to bestow an environmental ethic to a company by highlighting the environmental accountability of its employees.

Performing Skepticism

At the same time as engineers sometimes celebrated this intentional inter-mingling of personal and corporate values and accountabilities, they also took care to assert that the alignment was not the result of having been brainwashed by their employers or managers—in a sense, claiming their own agency rather than allowing an interpretation that would ascribe it to the corporate forms employing them.[30] Laura, a petroleum engineer, emphasized her own agency in critically assessing the company's outreach materials. She began the interview by positioning herself as coming from a "science background" that taught her to always question, "Where'd you come up with this? How do I know it's the truth?" She portrayed herself as initially skeptical of "corporate," which she referred to as *they*, distinguish-ing it as an entity separate from herself:

> When I first started the [outreach] program, I'm like, "Oh man, what are they asking me to do here, and how much do I really believe in it?" You know, the company's telling me to do this, but if you don't believe it yourself, it's useless. I mean you can't do it. And so I struggled with that at first. I'm like, "Is this just, you know, corporate trying to get me to do something and why am I doing this"?

As she described her growing "belief" in the company and the program, she switched to using the inclusive *we* instead of the separate *they*: "I kind of dragged my feet on it at first, but the more I got into [the program] and I started educating myself on what we're doing, how we're trying to mitigate, I'm like, 'Wow, we are really trying compared to other operators.' So there was that whole buy-in kind of perspective on a personal level that I had to go through."

Relating skepticism about corporate discourses and its eventual resolu-tion through analytic reasoning wove through engineers' narrations of their careers. Abigail said that though she had misgivings about work in the oil and gas industry, "I chose to work for [company] because they did things the right way. I wanted to be a part of that, a source of positive change from the inside." River, the petroleum engineer who lived in a town generally opposed to oil and gas production (chapter 2), said he came to welcome the skepticism that came from his left-leaning family and integrate it into his work in the industry. He said he kept a critical eye on what the company is claiming and tried to seek out other interpretations or viewpoints before arriving at conclusions. He explained, "I would always question if I was getting brainwashed by what the company was telling me. Am I missing a piece of the puzzle or a piece of the data?"

Through such confessions, engineers drew attention to their own agencies analyzing corporate discourses before adopting them, in effect asserting their own agency as distinct from the discourses and mandates of their corporate employers. Anthropologists would argue that these performances of skepticism also, however, reinforce corporate authority: they perform yet contain internal differences and portray companies as being willing to listen to detractors who think those corporations are in the wrong, yet they are ultimately offered as evidence of the company being in the right.[31]

Concealing Corporate Personhood

Not all of the engineers I met relished enacting the corporate person in their social circles like Kevin and Abigail did. Those who described how they concealed their corporate employment were careful to attribute their unease to what they perceived as narrow-mindedness among their friends and family, not any ethical quandaries with the companies employing them. River was married to fellow mid-career petroleum engineer Emma (whose memorable confrontation with landowners is analyzed in chapter 2), whom he met as an undergraduate. Both tried to minimize possibilities that they would be recognized in their small town as representatives of their companies or the oil and gas industry. They said they did not view their employers as ethically suspect or in conflict with their own morals but felt their own ethical commitments sometimes aligned more closely with those of their corporate employers than with those of their families. River was born and raised in the town where they lived. He described it as a "small, liberal, green community" and his mom and dad as "very, very green parents." He vividly remembered their disapproval of his career choice when he returned home to tell them he was going to major in petroleum engineering.

So when I came home . . . I had multiple members of my family and multiple family friends tell me what a terrible decision that was for a number of different reasons, everything from it's a diminishing resource and the industry is not going to be around long to, "This is a slap in the face, how could you?" And I had some family friends that are landowners within the productive portion of the [area] that were strongly telling me, "You need to come out, you need to see these wells, you need to see what this company has done to my land, and then you need to reflect on yourself to determine if that's what you really want."

After many conversations such as that one, River said he stopped trying to change his family's minds about his industry and resisted their attempts to position him as representing his employer, steering conversation instead to their common interests enjoying the outdoors. Indeed, it was River and Emma's shared love of the outdoors and family that pushed them to move back to River's hometown once they both found oil and gas jobs in the area. The town was popularly recognized as a mecca for outdoor recreation, including skiing, snowshoeing and other snow sports in the winter and hiking, camping, and biking in the summer. River and Emma enjoyed those activities with their two young daughters and an extended network of friends and family. While they shared their love for the outdoors with that social circle, they both kept quiet with them about their professional work, even though they both loved their careers.

Frustrated by how "close-minded" Emma found people to be about her work and her employer—a major multinational with long-term interests in the town—she began consciously concealing both her occupation and the company employing her when meeting new people.

I know [my company] is very involved in the community here but they try to run under the radar because here specifically it is not perceived very well. If [the company] is sponsoring something, it's not perceived as if they are doing a good job, it's perceived as if they are trying to make up for something they've done wrong. A number of years ago, all the [company] trucks had decals on them and . . . they had to remove all the decals from the trucks because people were getting harassed and trucks were getting vandalized. I feel like it's kind of sad, because [the company] is such a big part of the local economy and people don't talk about it. If I go someplace and I don't know who is there and I am asked, "Who do you work for?," I just say I am an engineer. *I don't say I am a petroleum engineer. I don't say I work for [the company] unless they push. We don't talk about it.* (emphasis added)

River and Emma were not alone in downplaying both their professional selves and corporate employers in their kin and social circles. A close friend of theirs shared that she stopped wearing a biking jersey emblazoned with the logo of the major oil multinational she worked for after fellow cyclists kept flipping her off while riding. In this case, her clothes made her available to be perceived as representing her corporate employer to other cyclists, many of whom take pride in reducing vehicular travel. While sociologists such as Kunda show that some employees create boundaries between their professional and personal lives to protect their personal sense of self against a hostile corporation, River, Emma, and the cyclist did so because they felt that their activities and ethical commitments as petroleum engineers were unwelcome in their kin and social circles. Their status as petroleum engineers who worked for oil and gas companies rested as one of their "potential identities" that could be made visible in their interactions with others.[32] To reduce this potential recognition, they minimized the information that others held about their professional lives.

RELATIONAL PERSONHOOD AND DISTRIBUTED AGENCY

Given the delicacy with which engineers understood and managed their participation in the corporate person, the anthropological concept of relational personhood provides a more nuanced theory of engineers' agencies in the context of corporate employment. Though the terms *self* and *person* are often used interchangeably, anthropologists tend to treat the self as the "subjective and experiential sense that one is or has a locus of awareness—a private consciousness that, while it may be a universal human trait, is also socioculturally mediated."[33] Personhood, in contrast, most often refers to a socially recognized and assigned status of being a social, embodied, and sentient being, which provides a window from which to understand "distinct ontological and ethnopsychological ideas about the constitution of persons, including persons' articulation with others, their interpenetration with the world around them, their moral or jural capacities, and the qualities of their agency."[34]

Through cross-cultural comparison, anthropologists firmly demonstrate that conceptions of "natural" persons as bounded, unique, self-aware, and autonomous individuals are not universal.[35] For example, Marilyn Strathern draws on Melanesian ethnography to propose the influential concept of a relational person who is the "plural and composite site of the relationships that produced them."[36] These "dividuals" are composed out of relations, specifically exchanges of bodily substances and circulating gifts. The ethnography in the following sections suggests that even though the engineers maintained bounded notions of themselves as individuals with particular histories and personalities, the corporate context of their work meant that they also experienced something more akin to an extended personhood, in which they were not always the authors of their own actions, found their own agencies expressed by others, and were held accountable for the actions of a distributed network of others.[37] If more relational notions of personhood help shed light on the corporate form, as Marina Welker shows, my ethnographic materials suggest that such notions are also good for thinking through the personhood of the people like engineers whose employment positions them to enact those forms.[38]

Almost all of engineers I met viewed organizational employment as inevitable for their profession. Being an engineer in a corporate context required them to submit to divisions of labor and institutional hierarchies that meant they were not always authors of their own actions. They had to take directions from their supervisors and their supervisors' supervisors. To do their "own" work, they had to use data created by others and work under constraints defined by others. They regularly had to pass their work on to others for revision, modification, and eventual implementation. Other employees and working groups inside their companies sometimes prioritized concerns differently, seeking to enact the corporate form to different ends. Anthropologist Thomas Yarrow vividly describes similar processes for architects, quoting novices who describe collaborative work by confessing that it is hard to "jump into someone else's brain . . . you have to swap your thinking to the other person's thinking." He identifies a "paradox of individual creativity: of a person that thinks and acts at once as themselves and through others."[39] This paradox was also present for my interlocutors, posing direct implications for the interlinked accountabilities of corporate forms and the people who embody them.

Exploring how my interlocutors navigated this distributed agency sheds light on how engineers themselves attuned their actions to the desires, mandates, and agencies of others, at the same time as they tried to influence the desires, mandates, and agencies of those others as they tried to enact corporate forms to more accountable ends. This analysis shows how employees tried to hold together the corporate person in the face of competing enactments and again points to the social license to operate (SLO) as a key but contested mechanism for attempting to align disparate agencies.

The distributed agency required by corporate work could be a source of professional angst. An environmental engineer who worked for one of the world's largest oil companies, for example, was disturbed by the lax environmental standards to which some of its global operations held themselves. He said, using the *we* of the company as a whole, "You're like, 'God, we're not really . . .'" and then trailed off, leaving a pause that spoke volumes. He continued, "When you look at those kinds of places sometimes, you see we're not really doing things the way we should be doing things, but there's no community demand either, right? So you have to do what you can to influence, to do better with local staff." Here the engineer used the *we* to signal the collective responsibility—and public accountability—for far-flung operations that did not meet his (and, he implies, the larger company's) standards for ethical performance. Faced with this mismatch, he tried to use his position to influence his colleagues to change their approach to align more with his own.

Kevin described a more positive but still contested attuning of his and coworkers' agencies. When I first met him, he was a reservoir engineer who supported the planning of new oil and gas fields by estimating reserves, conducting economic evaluation, designing completions strategies (the techniques used to make a well ready for production, including the fracturing techniques), and analyzing options for managing produced water (the water and its dissolved materials that flow upwell as a by-product of oil and gas production). He later was promoted to more of a planning role himself, where his responsibilities included pad size optimization, infrastructure planning, and process management. He made recommendations for how far apart the wells should be spaced, how long they should extend laterally, and how they should be fracked. Engaging in all of this work required him to manage, influence, and adapt to other agencies.

Kevin said he respected his boss, a fellow but more senior engineer, but described himself as "annoyed" by how the boss factored "intangibles" into the decisions he made and then forced the team to make. Their responsibility as reservoir engineers was to estimate and protect future value, including their company's ability to drill new wells and entire new oil and gas fields. They were brought into wider discussions that would bear on that ability, such as whether and when to plug and abandon existing wells. When a well's production dropped off, they could direct other employees either to use engineering techniques to stimulate more production or to end its productive life by cutting it off below the surface and burying it. Kevin and his coworkers used data sets on a well's historic production and the performance of comparable wells in the field to estimate how much more the well could produce and assign a dollar figure to the remainder using market projections. They could then compare that figure with the costs and benefits of the "stimulation techniques" to enhance production. Where things became murky was factoring in the social ramifications of their decisions. Kevin explained, using the *we* of his small team:

> A lot of it is looking at future value of the field. So if you have a landowner who owns, who is the gateway between us being able to drill a well or a pad of wells a lot of times, you look at your estimated recovery from a new drill, and that typically tips the scale. So we say, "Well, this well is still valuable, it's worth half a million dollars. But compared to our potential to drill here sometime in the next five to ten years, it's better to get rid of this well now." And so that was usually the trump card we'd play when it came to these discussions, the ability to drill a well in the future. . . . So you have to pick sort of this value threshold, but also factor in the value of improving your relationship with that surface owner.

Quantifying the value of a landowner's acquiescence resonated with how Kevin described his decision making in general by saying, "I always think there's a dollar figure you can put on everything that you do." This style of analysis is a form of commensuration characteristic of engineering practice.[40] For Kevin, this approach resonated with engineering problem solving, which had clear comparisons, translatable units, and one right answer. In contrast, he said, his boss would make decisions based on "gut feelings" even if there was "no dollar amount that tells you to do it at that point." He and his boss therefore had to compromise, with the hope of "agreeing on a decision for different reasons."

Plug-and-abandon decisions also revealed tensions among teams, such as the landmen, who negotiated the lease agreements with mineral and surface owners, and the production engineers, who were responsible for protecting the current value of wells more so than the future value protected by Kevin's team. The landmen's desire to ensure positive relationships with current landowners and to nurture community-wide acceptance meant that they preferred to plug and abandon wells that caused friction with landowners and nearby residents, even if the production engineers could extend their life by bringing in large amounts of equipment to workover the well. This created what Kevin called a fault line between "people who are interacting directly with the owners versus the people who are in charge of the decision of the value and protecting the value to the company." Because even his fellow coworkers were accountable to different supervisors and work teams with different expectations—to say nothing of their varying personal ethical frameworks—Kevin had to learn to attune his own actions to others while also trying to shape the agencies of those others to align more closely with his own.

SOCIAL LICENSE AS AN ALIGNING DEVICE

A key device for facilitating the alignment of otherwise disparate agencies was the social license to operate (SLO). Chapter 2 proposes that the SLO was prevalent in the engineers' narratives, as well as official corporate discourses, because it purported an alignment of industry actors' accountabilities to formal standards and policies, to their profession, to the public, and to their personal ethical frameworks. Chapter 3 provides a historical account of how the concept privileged industry by framing questions of resource production to be those of *how* such production can be done responsibly rather than *if* such production should happen at all. Those chapters primarily focus on relationships between corporate actors and their multiple publics. In this section, I illustrate how the SLO also served a crucial internal purpose of attempting to align the disparate agencies characterizing the corporate form.

Kevin explained that different teams inside of his company "can be at odds, but everyone's job is to drive value, so you have to kind of come to agreement at some point." Though the word value can have multiple valences, Kevin privileged the economic sense. In the framework he described, social acceptance figured as an organizing device to align employees and teams with disparate work responsibilities. Kevin explained:

> One person's goal is to drill the most wells. One person's is to drill them cheaper. All the tension is typically because their goals aren't the same. But from a stakeholder standpoint, I feel like I've never seen a conflict where people don't pick something that benefits our stakeholders. I think the only tension is . . . just defining that line between making a decision that benefits the stakeholder versus retaining the value where we have the right.

While Welker compellingly shows that the rhetoric of the business case can be used by corporate personnel to justify multiple and contradictory ends, Kevin's experience working in the midst of heated debates about fracking also speaks to the power of the SLO concept to bring together teams and employees with otherwise disparate interests. This power stems from its grounding in a business case that proposes community acceptance as profitable and profit as the ultimate goal of a company, as described in chapter 3.

One of the key differences between how mining companies such as AMAX Minerals Inc. and oil and gas companies such as the one employing Kevin was the dispersed nature of oil and gas production. Whereas mines are spatially intensive, their footprint does not have change substantially.

The oil and gas wells operated by Kevin's company, in contrast, were interspersed with suburban developments and ranches, and the company needed to drill new wells constantly to remain profitable.[41] This meant that the number of the company's stakeholders was also constantly expanding, opening up opportunities for the company's image to be sullied. In Kevin's words, the ability of landowners to "change your reputation in the community is probably the biggest reason to work with them, because if we ever want to drill new wells or if we want to improve old wells, then you have to have the community see what you're doing as a good thing for them to say yes." Kevin saw that this line of reasoning was compelling to stockholders each time he was tasked with joining a quarterly conference call.

> And so the cool thing about [the company] is that we have our stockholders on the same page as us from an engagement standpoint, because they will ask us questions about our ability to change the conversation in places like Colorado because they know that the value of their stocks that they hold is dependent on our ability to do things the right way in Colorado so that people have a good impression of us. These people, these bankers that seem so anonymous and just driven by profit are concerned that if we do things the wrong way in Colorado, then it hurts their ability to maintain the value of the portfolio.

Kevin perceived his company's shareholders not as an anonymous, aggregate mass but as specific people who had values that encompassed doing oil and gas development the "right way." Whereas social scientists may argue that such values-based investing ultimately encompasses moral concerns under the economic desire for profit,[42] Kevin portrayed economic and moral value as standing on equal footing. He argued that what held the company, its leadership, and its investors together was a marriage between what Kevin and his coworkers described as "theology and capitalism." The company was led by a charismatic CEO who publicly described himself as a "Midwestern Christian" and enjoined his employees to view "work as a spiritual enterprise."[43] For Kevin, this meant that his company attracted stockholders and employees who shared a common set of values, making it easier to enact them on an everyday basis. He contrasted this with other companies by saying:

If you're a publicly traded company, you have this weird allegiance to an anonymous stockholder that can skew how you think about things. Everyone I talk to, any industry, you're going to have that conflict, and it may, even if you're not publicly traded, it's your private equity owners versus your leadership versus you. But, I think where I'm at, they all line up together, which is really cool. And it makes working, makes the drive to work so much easier.

Kevin described the alignment between his values and those ascribed to the company as a source of professional fulfillment. He said the company's values were the main reason he chose his company when making his first job applications. He described the people he interviewed and eventually worked with as "people that lived within my own personal values and the corporate values that we have were actually lived out, and they aligned." He spoke directly to the unique personhood of corporations, saying, "There's a human side to everything that we do, and we can't separate the people's individual morality from what it feels like as a company we're trying to pursue." He then extrapolated a person's behavior to representing the industry as a whole: "You're also thinking about if you do things wrong, then you're affecting the industry's reputation as a whole, which is an even bigger burden to bear." The alignment Kevin felt between his company's leadership, investors, and employees formed the basis of his desire to work for the same company for the rest of his career. Kevin could have made a variety of arguments for the importance of this alignment, including prominently that such values would translate into steadier economic growth that would provide job security in the face of perpetual booms and busts. But instead, Kevin invoked the soul, saying, "I think there's tangible value to your soul to do things moral."

Kevin's comment brings us squarely back to the persistent concern that corporate work compromises engineers' souls, as influentially articulated by Layton and reiterated again and again by engineering educators and critics. As much as Kevin positively enacted the corporate form in our interview, even he was not a one-dimensional "organization man" of the kind critiqued by midcentury sociologists. In narrating his career, he acknowledged that corporate work could compromise an engineer's soul.

He experienced stresses and frustrations on the job and did not always agree with the decisions made by those around him. Yet he stayed, buoyed by an overall sense that enacting the corporation did not require him to compromise his own values, as that multiple, heterogeneous corporate form was held together by values that resonated with his own: by treating stakeholders well, the company should profit.

DETACHMENT

Whereas the engineers profiled in the chapters so far found a way to stay in industry, others left. Some became consultants, hoping to find more professional autonomy, as analyzed in chapter 5. Those who left industry entirely all described feeling constrained by the kinds of questions they could ask in their position as corporate employees. Elijah left a career in oil and gas seeking what he called "big R" rather than "little r" responsibility. He was a geological engineer in his twenties who had cultivated his own philosophy of accountability through what he described as a countercultural childhood in the American West, where he learned what he called the "intrinsic value and rights of nature" as something more than a resource to be developed. He further developed this view of the natural world when he started rock climbing as an engineering undergraduate student. After graduating, he went to work at a medium-sized oil and gas firm in Houston because few other options were available and he had student loans to repay. The company where he worked had an industry reputation for a forward-looking approach to environmental and social performance. He pushed the envelope by conspicuously reading books that were critical of fracking in the breakroom while he ate his lunch. His coworkers never criticized him, he said, seeming to respect his desire to understand their opponents. But he felt that the frameworks of responsibility presented to them by the company leaders was not robust enough to encompass the questions he wanted to ask and the discussions he wanted to have.

The company's framework, grounded in the SLO, addressed only what Elijah called "little r responsibility." He mouthed the dominant discourse by explaining, "If we have an environmental disaster, there would be some serious ramifications from the state government or the EPA or the public. . . . The last thing we want to do is harm any social license to operate because that could overall impact production and profits." He halfheartedly described himself as initially being "fine" with their approach but ultimately frustrated by how it constrained the questions he could ask about society's use of natural resources. Turning to use the *they* instead of the *we* of the company, signaling greater distance between him and his coworkers and supervisors, he said:

> With the little policies that they did, I actually felt pretty comfortable. They made sure to design their wells with a pretty high safety standard. People who disregarded these standards were severely reprimanded or terminated. I was pleased with those details. But to draw that distinction—I felt uncomfortable in Houston, a city designed for and by petroleum that is this energy draining, pinnacle of the problem of how our cities are designed. So I felt comfortable with my company's ability to stand out from other companies in the social responsibility mythology in their operations. But at the end of the day, I felt uncomfortable with the fact that I was living in a city driving over twenty-five miles to the office every day in my car. I ended up using a van pool when they finally got that set up. But you can imagine that if you have this broader philosophy, which I had, there's only so many microexcuses you can make to yourself until you find it impossible to reconcile how you're living with what it is that you're doing.

What Elijah called "big R" responsibility entailed asking probing sociotechnical questions about how society was structured, not about how a particular well was designed. Those "big R" questions interrupted the underlying vision of complementarity animating the SLO framework for thinking about corporate accountabilities because they questioned the need to continue producing the same quantities of oil and gas in the first place. Lacking space to ask those kinds of questions at the company, he began working internationally for Engineers Without Borders before returning to school for a graduate degree, hoping that it would allow him greater opportunity to find work that allowed him to ask and answer more fundamental questions about natural resources and society.[44]

Sofia, originally trained as a civil engineer, also returned to graduate school after becoming disillusioned with her corporate job early in her career. Of all my interlocutors, she was the most critical of the reluctance of mining companies to fundamentally rethink their approach to communities. Sofia had long dreamed about working for the big mining multinational that operated a large mine close to where she grew up. When she eventually secured an engineering job with the company, she found herself awkwardly placed between the mine personnel—almost none of whom came from that region—and the poor communities who lived near the mine. Using the term *they* to describe her former coworkers and signal her own distance from the corporate form, Sofia said:

> They [the company employees] are not really listening to them [the community members]. They're not really empowering them. They're not allowing them the opportunity for self-determination. They just put them down every time they meet with the community. They practically almost laugh in their faces, you know, because their concerns are different than theirs. That's why I walked out of there [the company]. The biggest thing that I took away was that, that it's not a public relations issue. It's something that was inherently embedded within the entire system of whose knowledge is valued and whose comments are appreciated and whose aren't. And sometimes it may—I don't know—it may have really been unfounded complaints. But the fact that they continue to approach them from such a disdain kind of perspective just made that even worse. Like it automatically, you know, just stopped any bridge for communication and the willingness of anybody to engage in a constructive session.

Sofia critiqued the systematic devaluing of local knowledge by the company's coworkers and managers. In her view, they espoused a commitment to listen but did not respect differences of experience, concern, and knowledge.[45] They did not rethink their own assumptions or even act on the concerns that local people brought to them. This disconnect made it impossible for her to invest her professional identity in the corporate form employing her, even though she had strongly desired to work for them because of the power they wielded in her home region. Her questions about knowledge and power led her to pursue a PhD that would allow her to ask and answer the kinds of questions that animated her broader concerns about social justice. At the time of our interview, she dreamed about bringing engineering students to community meetings to help them "debunk" the elitism on the part of the company engineers responding to criticism and questions from the public.

DETACHMENT AND GENDER DISCRIMINATION

Workplace gender politics played a significant role in pushing out women engineers who had broader aspirations for industry accountability than they found in the corporate world.[46] The mining and oil and gas industries are frequently viewed as archetypes of masculinist work and male-dominated organizations. In their study of women geoscientists working in the oil and gas industry, sociologist Christine L. Williams and her colleagues found that corporate diversity programs can paradoxically reinforce male dominance and that the shifts in organizational labor glossed as the "new economy"—job insecurity, teamwork, career mapping, and networking—reproduce gender inequality.[47] In my own prior research, I grounded the relatively successful integration of women engineers and technicians into the Wyoming coal industry in the particular regional history and cultural context that established more gender-neutral expectations for work in the industry.[48]

For the women engineers I met who left industry, they attributed their decision to two sources: their unfulfilled desires to ask different kinds of questions and the gender discrimination they experienced. Addie, a metallurgical engineer at the beginning of her career, left a corporate job in mining after encountering resistance to her conviction that natural resource production could not continue increasing exponentially. Her departure was striking, given that the ethic of material provisioning originally inspired her to work in mining. She remembered:

> I could really get behind this idea that every day I made a product that went directly into a stream of more products. So literally, everyone was impacted by the work that I did: anyone who goes to a hospital, anyone who picks up the phone, anyone who has a roof over their head or copper pipes in their home. I could really, deeply feel that the work that I was doing was going to have an impact. I could literally see where it was.

Yet within a few months of beginning her career at one of the world's largest mining companies, Addie realized that its executives were resistant to considering alternative solutions around mining waste and recycling. She took part in the company's environmental and social outreach activities, such as river cleanups and elementary school visits, but began to feel a powerful disconnect between her environmental commitments and those that held together the rest of the company. "I could see that what I was doing like in theory was good," she recalled, "but it was really hard for me to feel that those engagements were actually authentic because I was basically just telling people, 'Keep buying things that have all of these goods in them because I'll go to the river and pick them up when you've dumped them there.'"

Addie also routinely experienced gender discrimination and sexual harassment at work. She reported having a boss who told her that "he didn't know how to speak with women" and "specifically started hiring only men" when he had multiple females reporting to him and he "felt outnumbered." She recalled instances in which women's advice to advance the position of women inside of the company was solicited and then flatly ignored. She summarized, "It was overall very hard to be a female. It's hard to progress as a female." She said, "The personal social side of things is really what ultimately made me leave. . . . I was never going to be able to make it through that organization and not get chewed up and spit out." She quit after three years to pursue an independent career in social impact consulting for organizations with no direct ties with the mining industry.

Other women left corporate work because they experienced discrimination, harassment, and thin corporate accountability but maintained a foothold in industry as consultants. Diane, a senior metallurgical engineer, experienced gender discrimination at her corporate job and could not stomach the unethical metallurgical accounting practices she witnessed. She recalled, "A friend of mine actually came to do due diligence [at the company] and I realized I didn't even want to invite him over for dinner because I didn't want to have a conversation about what was happening. It was like ethically I didn't think it was correct. I didn't want to be a part of it." Diane made her disapproval known while she was leaving, saying tersely, "I very seldom go quietly without people understanding. . . . My philosophy is that I will stick around if I think I can change something, but when I reach a point that I understand that it's not going to change, it's time for me to go do something else."

June, a thirty-year mining engineer, left corporate work after tiring of constantly trying to convince her male colleagues to follow her suggestions to integrate social acceptance into their decision making. Their insults of her expertise prompted her to start carrying a coffee mug from her alma mater, the Colorado School of Mines, into her meetings: "I'd set it on the table because that was the subtle way of saying, 'Yes, I know what the hell I'm doing.'" She described being passed over for promotions and being paid less than men with similar or fewer credentials for doing the same job. She darkly joked about the stakes of reporting and attempting to rectify such inequities, saying, "I didn't push the sexual discrimination lawsuit because if you punch that one, you have to win the lottery and make everything you want because you're not working again."

June took particular pride in integrating local social concerns into her work of pit design and planning but frequently became frustrated when other engineers did not because they took a more narrow view of profitability. When designing a potential future mine, she thought holistically about the political, economic, and cultural context as constraints on her work. She described a project in South America in which the original open pit design would have taken out a rock formation with deep cultural significance for the local community. She redid the design to leave the formation in place.[49] Invoking the importance of the SLO, she successfully persuaded the company decision makers to follow her recommendation, even though it would "cost" the company several million dollars of potential profit. "But the key is once you piss off the community, you're never getting it back. You have to have your social license to operate. You either can [maintain the SLO] or you have no choice and you're done." On the heels of positive buzz about the project, the junior company was bought out by a large multinational. According to June, its project manager tossed out the pit design and the stakeholder engagement efforts she and her colleagues had developed, which turned the community permanently against the project and prohibited it from moving beyond the exploration stage.[50]

Even though June's disappointment in the project not coming to fruition was palpable in the way she spoke, she did not begrudge the community for halting a project that would have been devastating to them. She said, "The interesting thing is in all of our careers, the one thing you have to do is look at yourself in the mirror. I do that on the social responsibility, on my engineering, on my ethics, and what I did at the job that day." She maintained some hope that engineers would eventually critically assess their projects from the perspective of the people who had to live next to them, if they could generate some empathy for those communities. Speaking with the *we* of engineers who work in the mining industry, she said, "The biggest problem we have is we've got to get out of our own skin. We may not understand their ways of life or anything else, but we have to stop and look at it."

*   *   *

June felt that consulting made it possible to reconcile her accountabilities to herself, to her profession, and to the public with the projects she chose to work on—a common theme explored in greater depth in chapter 5. In contrast, Addie, Elijah, and Sofia left their industries entirely because they did not find space with the corporate forms employing them to ask the big questions that motivated them. Rather than taking continued natural resource production for granted, Addie and Elijah wanted to rethink ways to reduce consumption. Rather than take hierarchical relationships between mining companies and marginalized communities for granted, Sofia wanted to rethink systems of knowledge to empower the people disadvantaged by the current ones. Both sets of questions exceed the frameworks of accountability that their corporate employers and industries in general presented to them. Addie's and Elijah's questions directly challenged the ethic of material provisioning by calling into question the need for such high levels of continued resource production. In so doing, they also challenged the company's limited identification of their own public accountability: to produce minerals, oil, and gas in an environmentally sound way. Sofia's questions challenged the authority of the corporate form to define the terms of their public accountability in ways that empowered themselves to continue mining while disempowering the nearby community to have their concerns addressed.

While Addie, Elijah, and Sofia left industrial work entirely, most of the engineers I met who were critical of the limitations placed on them by corporate employment were like June and became consultants. As chapter 5 argues, this different institutional location opened up more opportunities for them to choose corporate clients and projects rather than be assigned to them. Like Addie, Elijah, and Sofia, they sought to ask bigger questions about natural resource production. But unlike Elijah and Sofia, they valued being a positive source of change from the inside.

CONCLUSION

When they thought about themselves and their work, none of the engineers I met held illusions of themselves as completely autonomous agents, solely responsible for their actions. Instead, they acknowledged that their work in a corporate context entailed—at least partially—becoming part of a larger entity whose actions and impacts exceeded their own. By accepting corporate employment, they stepped into an extended corporate person characterized by distributed agency: they were not always authors of their own actions, and they became partially accountable for the decisions and actions of their coworkers, managers, and fellow industry actors in general. They were frequently called to stand in for an entire company or industry: family members, friends, and even passersby who noticed a corporate logo on their clothing or vehicles could hail them as enacting their corporate employers and therefore as complicit in their behavior.

Some of my interlocutors seemed to relish performing a meshing of their own ethics and the corporate "person," which entailed taking accountability for the actions that were collectively attributed to the corporate form and lending their own ethical commitments—such as to environmental conservation—to the corporate person. Others could not bear their agencies and attendant accountabilities being tied up with those of their coworkers, supervisors, and industry actors in general, prompting them to detach from the corporate forms employing them. Most, however, described their experiences falling somewhere in the middle. When using *we* to refer to the corporate form, they emphasized their own enmeshment of the agencies and accountabilities of its disparate parts. When using *they*, they signaled that they were standing apart from that larger collective, almost always because they could not abide the actions undertaken in its name.

What does all of this suggest for Layton's fear of engineers sacrificing their souls for corporate employment, and its continued reverberations among engineers and the people who study and teach them? The engineers I met who stayed in industry felt that corporate work sometimes expressed many elements not just of their sense of self but of their *souls*—their understanding of what was good or right and their own efforts to bring it about. Enacting a corporate form did not always or automatically mean disowning that soul. *They themselves* were the corporation—at least part of it, and for some of the time—and they tried to align its disparate parts around their own ethical frameworks of what was good or right. Faced with managing their participation in an extended person comprising multiple agencies beyond their own, they appealed to concepts such as the SLO or company policies committing to environmental and social responsibility to influence agencies other than their own.

This relational view of the corporate form and the distributed agencies comprising it proposes a different approach to engineering ethics. Anthropologists keenly note that distinctions between the "relational" Melanesian person and the autonomous Western individual are likely overdrawn, as multiple notions of personhood can coexist in a single place, and the stereotypical Western individual is an idealized legal construct recognized as better reflecting philosophical traditions than actual human experience.[51] Yet it is precisely this idealized, liberal autonomous agent that forms the backbone of US undergraduate engineering ethics education—a field dominated by philosophers trained in Western traditions.[52] In short, an unacknowledged theory of personhood and agency underpins US engineering ethics training that is at odds with how the engineers I came to know thought about the responsibilities and accountabilities of their work. My interlocutors did not just encounter corporations as behemoth external forces that tried to turn them into relentless profit maximizers, requiring them to either acquiesce by becoming automatons or resist by becoming whistleblowers. Instead, they themselves enacted the corporation, even though they could not fully control the rest of the agencies constituting this form. When weighing decisions about potential courses of action, they did so from a position as one node in a complex network in which they were accountable to and influenced by others, not as autonomous agents.

These ethnographic materials pose new questions for engineering ethics, especially relating to corporate employment. What does "professional autonomy," construed as the prerequisite for public accountability, look like in a model of personhood as relational and extended? Whereas whistleblowing figures engineers as autonomous agents who either act against a separate corporate form or are completely subsumed by it, we could instead consider how engineers themselves enact corporations to different ends and under different constraints. Rather than simply celebrating the clear "resistors" who challenge and then leave their corporate jobs, my interlocutors invite us to consider the agencies involved in improving relationships with landowners through plug-and-abandon decisions or in designing more sustainable mine plans and infrastructure with ecologists. While these may be critiqued by skeptics as projects of reform rather than revolution, they are important areas of practice that explicitly recognize engineering as a sociotechnical phenomenon and seek to make industry more accountable to its publics.

The engineers profiled in this chapter all positioned themselves as having found at least some peace with the corporate context of their work, pointing to moments or spaces in which they had reconciled—at least partially, and at least some of the time—their sense of self and profession with the corporate person they were called to enact as employees. Given the context of their work at the time of our research interactions, the viewpoints they expressed in interviews and conversations emphasized harmony over conflict. In contrast, my interlocutors who either left or did not pursue corporate employment opened up more space in our research interactions to critique organizational work. Most had created small, independent consulting firms that allowed them more professional autonomy to choose the companies they worked for and the projects they would accept. Chapter 5 turns to their experiences in detail to explore how they fulfilled their passions for their work in industry while setting bounds on the activities for which they would be held accountable.

5 LIMITATIONS OF LIMINALITY

Gary was an exploration geologist who hoped he would never make a big find. I met him while he was transitioning into retirement in his sixties and building support for what he called a "new way for geology to serve society." We traced his story back to his childhood, when he became enchanted with rocks, fossils, and the outdoors while growing up in the American Midwest, where he fondly remembered hiking with his family through trails that were strewn with fossils and geodes. Upon graduating from an engineering college in the late 1970s, he found that the only people hiring were mining and oil companies. He worked as an independent contractor for a string of multinationals, "niche-hopping," he said, from "one commodity to the next." After spending a few years looking for mineral deposits that could become the basis for new mines, he then moved to oil-related mapping, remote sensing, and air photo interpretation for major oil companies. Yet he approached this work with a persistent sense of trepidation that, if he actually made a discovery, it would lead to a major industrial development that would wreak havoc on the environment he loved so much: "I was always afraid of finding something. That was because I love the outdoors. I've always been a camper and a hiker. I didn't want to see anything messed up. I would bet that a lot of the geologists, the people that go into this business and find themselves in exploration, have that same regret. They're afraid of actually being successful in the corporate world."

Gary experienced a profound sense of alienation in his work as a corporate consultant. He loved that exploration gave him the chance to piece together a geologic picture from field observations, but he eventually realized that his goal was never the same as his employers'. Reflecting back on his early career, he said that while he was occupied raising a family and taking care of others, he could not see that disconnect fully or imagine an alternate career path that would offer an "idyllic job that actually paid." Describing how he dealt with the anguish of thinking that his work would facilitate large-scale environmental destruction and culture change, he said, "You just put it out of your head. You just don't think about it, you compartmentalize. You join the Nature Conservancy and then you go out and look for gold." In remembering his corporate days, Gary described a process of detachment from the broader context and effects of his work—"you just put it out of your head"—that produced a compartmentalization of where and when he could invest himself in his work. He remembered feeling estranged from the corporate firms that contracted him: "You end up having this black cloud that follows you around. You're never really accepted. You're always one of those contract people that they can have do anything anytime. They don't worry about you. There's no loyalty there, but then again, there isn't in most of corporate America anyway. It's all gone." The lack of loyalty Gary perceived in corporations was one part of his larger critique of them. As I explain later in this chapter, Gary went on to found his own nonprofit to find ways to make mineral development benefit the poorest people living in close proximity to these resources.

While Gary's critiques of corporations were fervent, they were not unique in my research. The engineers and applied scientists I met who had left corporate jobs described a persistent and untenable clash between the expectations they felt as company employees who reported up a chain of command, on the one hand, and their own accountabilities to their professional ideals, to the publics their work affected, and to their moral frameworks, on the other.[1] The lure of consulting was that it seemed to promise greater opportunity to choose clients and projects that already resonated with their other accountabilities. They narrated their careers as though they held their professional, public, and personal accountabilities constant, treating the corporate projects and clients as variables that could be selected and adjusted to fit the prior three.[2]

This chapter shows that one of the biggest challenges facing these consulting engineers and scientists was managing a liminal position in relation to those corporate forms. This liminal position sharpened their experience of distributed agency and positioned all but the most successful in a position of dependency in relation to the corporate clients they "chose." In a different but related context, Javiera Barandiarán evocatively quotes a young environmental consultant in Chile who described the relationship among scientists, consultants, and companies as "toxic" because of the power of the corporate actors over the others: "Consulting firms are in a vicious circle because the company pays you to do a study to evaluate the company's project's environmental impacts. The company is judge and jury in its own cause."[3] I heard similar echoes in my own interviews. In particular, consultants felt hamstrung by their position "recommending" courses of action to clients that then had the power to act—or not—on their work. While consultants enjoyed relative autonomy in designing and conducting studies or creating plans within the client's specifications, they were not in control of how those were eventually implemented by the people hiring them. I conclude by suggesting that the bind in which many consultants found themselves ultimately served as a source of value and legitimacy for their corporate clients: consultants were often perceived by publics as more "objective" and "independent" than company employees, even though in practice they remained financially dependent on those companies for their livelihood and did not retain control over the implementation of their work.

CORPORATE DETACHMENTS

This chapter theorizes practices of detachment from corporate forms as a central dimension of engineers' agency. Detachment is a key dynamic in everyday life, though anthropologists' own cultural predilections for connection and attachment lead many in the field to foreground these latter processes over the former.[4] Detachment figures in the anthropological literature on corporate social responsibility as a corporate strategy for separating these entities "legally, morally, and socially from binding obligations and responsibility to producers."[5] This process is vividly illustrated through

Dinah Rajak's ethnography of mines strategically delimiting the beneficiaries of their programs to a "working community" that does not encompass subcontractors or miners' own families.[6] It is at work in the research by Katy Gardner and colleagues on the "community engagement" practices of a transnational oil company that actually further separate it from the community through its grounding in neoliberal values of "self-reliance, entrepreneurship, and "helping people to help themselves."[7] It is present in Elana Shever's analysis of a transnational oil company that abdicated its responsibilities for a health center in a poor community through a corporate social responsibility program (unironically) titled "Creating Bonds."[8] And it animates Hannah Appel's ethnography of offshore oil enclaves in western Africa that prompt companies to work furiously to "perform a distinction between itself and that which is 'outside' its walls, despite their utter intercalation."[9] Jamie Cross argues that "the bracketing, limiting, and ending of economic relationships, like those between actors in a market transaction, are always still relationships; and, to the extent that detachment is a guide to conduct, it is an ethic."[10]

Taking up the question of engineers in particular, Penny Harvey and Hannah Knox argue that their practice is predicated on "virtuous detachments" in which engineers extract themselves and their work from ongoing sociomaterial relationships, such as the mutual imbrication of construction sites and the people and environments transformed by them. Importantly, they identify these detachments as political in nature, since they define the "practices for which engineers might legitimately be held responsible."[11] In my research, engineers detached from—but also partially reattached to—corporate forms precisely to set the contours for the accountabilities to which they would be held. This chapter builds on Harvey and Knox's work by drawing out the politics of the liminal status occupied by engineering consultants. They could not just detach but remained dependent on corporations for continued contracts and were frequently called to speak on behalf of those corporate clients in public to help diffuse social unrest.

Consultants move inside and outside of the porous corporate form.[12] During my research, the knowledge they produced became materialized in the infrastructure, processes, and discourses attributed to the corporations contracting them. In meetings with external "stakeholders" or public hearings, consultants were sometimes called to speak on behalf of those corporate forms. At other times, their status as nonpermanent employees was marked, such as when they were introduced as "independent" experts to shore up public support for proposals being made on behalf of corporations.

The position of consultants clearly illustrates the partible and permeable dimensions of the corporate person analyzed in chapter 4.[13] "Independent" consultants do an astounding amount of the work attributed to major corporations, for several reasons. While most of my interlocutors narrated their move to consulting as an agentic *choice* that they welcomed compared with full-time corporate employment, the job market also changed the opportunities available to them. Corporations themselves became leaner by cutting the ranks of employees for whom they owed benefits such as health care or retirement plans, and some specialized forms of expertise were required only at occasional points of a project, making it expensive to keep such professionals in house on a full time basis. "As firms become less self-sufficient, their boundaries become more permeable because lean firms must, by definition, acquire more resources externally."[14] The final environmental report put together in 1981 for the proposed Mt. Emmons mine analyzed in chapter 3, for example, lists only fourteen employees of AMAX Minerals Inc., along with fifty-six consultants working for sixteen different firms who participated in the studies included in the report. The number of consultants engaged would have been even larger if the mine had actually been fully designed for construction and then built. By 2018, a geological engineering consultant and project manager said that, by the time her firm finished with a proposal for a mine tailings dam, "the entire team that touched the project, I mean, it would be hundreds of people, eventually."

The consultants I met worked primarily as engineering and applied science consultants, though a few were engineers who moved into community relations and performance standard reporting. Their work departs substantially from the dominant image of consultants cast in the mold of management consultants. Felix Stein's ethnography of these professionals shows how they enable managerial rule and advance shareholder value through their work reconfiguring corporate relationships and processes.

He argues that these management consultants delivered not concrete products or knowledge but an image of legitimacy for the clients contracting them. Indeed, he writes that "almost all management consultants I met were somewhat proud of not being actual experts at anything."[15] In contrast, those I came to know had very deep and very concrete sources of expertise that were highly and specifically valued by their clients. Rather than producing endless PowerPoint slides unmoored from clear referents in the world, as did the managerial consultants, the engineers and applied scientists also produced studies, data sets, and plans that could be measured against the world. What they shared with the management consultants was their dependence on corporate clients for work and a liminal status as both company insider and outsider.

Consultants' participation in a corporate person understood relationally, as a composite of multiple agencies, made them complicit in the activities of many others whom they did not control and sometimes could not even influence. The difference between the consulting engineers profiled in this chapter and the full-time corporate employees profiled in chapter 4 is that consultants were institutionally positioned to have more opportunities to detach their activities, emotions, and sense of self from the corporate forms hiring them. While they did have to construct alignments strategically with the corporate forms that hired them, they were also able to stand more firmly outside of them.[16]

PROFESSIONAL AUTONOMY

This ability to stand at arm's length from corporate mandates has been productively theorized in terms of "professional autonomy." Sociologists and historians suggest that engineering work is in many ways necessarily bound up in organizational, if not corporate, settings: "Engineers, in order to function as engineers, must have a boss, or at least a client."[17] This feature of their work distinguishes engineering from other professions, such as medicine and law, and has generated much consternation for engineers and those who study them. In Edwin T. Layton Jr.'s foundational history of the US engineering profession, he writes, "The very essence of professionalism lies in not taking orders from an employer."[18] I join other scholars who urge a complicating of this account by paying attention to the different dimensions of autonomy and how these may be supported or constrained inside of organizational workplaces, rather than beginning from a point that presumes professionalism and corporate employment to be antithetical.[19]

My interlocutors seemed to accept organizational work as an inevitable requirement of working as an engineer but appreciated that consulting allowed them to stand at a greater distance from corporate forms. Sociologists Stephen R. Barley and Gideon Kunda found similar sentiments in their in-depth study of technical consultants, who desired independence from the "politics, incompetence, and inequities of organizational life."[20] They write in a celebratory tone, "Nearly all of our informants told us that contracting had released them from the social constraints of organizational life. They no longer had to conform to the whims of managers who once controlled their fate."[21]

My interlocutors found their tethers to corporate life to be much shorter, since they remained dependent on particular people and units inside of corporations for continued work contracts.[22] Any "independence" they found stemmed from their strategic detaching from and connecting to corporate forms in search of finding corporate clients and projects that aligned with their professional, public, and personal accountabilities. This also contrasts with Barley and Kunda's findings that consultants "could speak their minds while focusing more exclusively on the technical aspects of their work."[23] My interlocutors did not use consulting to retreat into a depoliticized technical world, quite the opposite: consulting work provided them more space to practice engineering that articulated with broader social and environmental accountabilities.

IN SEARCH OF ACCOUNTABILITY

Most of the consultants I met had moved into that work after becoming disenchanted with corporate employers, like Colton, a midcareer petroleum engineer who agreed to be interviewed with the caveat that if it fell on a "powder day," with new snow, he would be out of the office backcountry skiing. He was born and raised in rural Colorado, where he grew up loving the outdoors and helping his dad out on the oil patch. Unlike most of his peers, he paid his own way through college, which meant that he accumulated a lot of first-hand oilfield work experience. After learning that he preferred smaller companies to larger ones, and deciding that he wanted to stay in Colorado instead of getting "stuck in Houston," he started his own consulting firm that allowed him to "decouple," in his words, from the misdeeds of the larger industry.

Colton specialized in the design and implementation of difficult wells, such as high-pressure, high-temperature deep wells and shallow reservoir wells whose small economic margins made every design decision count. Colton took pride in his work, believing that oil and gas development could be done in a way that did not offend people who sought to use the same public lands for recreation, as he did (see figure 5.1 for an example of this style in general, but not Colton's projects in particular). He explained:

Figure 5.1
The equipment on this well pad in New Mexico is painted to blend in with the surrounding area to reduce the visual impact. Photo courtesy Bureau of Land Management.

I work on a lot of public lands, out in the West. I operate mainly in the Rockies, and most of it's public BLM [Bureau of Land Management] land. So I'm very conscious about the appearance of my operations. The BLM, they really like our operation. The state inspectors do. We keep everything buttoned up and looking good. We want to keep it a certain way, but it's driven by perception. You don't want to get a bad perception, because then that makes them look at you closer, right? I take pride in my operation, and I don't want it looking, I don't want somebody out there in the back country, hiking around, and going by my well and saying, "Well that's ugly."

Colton hoped that his own work would eventually help change people's minds about the possibility for responsible oil and gas development. Living in a Colorado town known as an outdoor recreation mecca and hub of progressive politics, he felt demonized on a daily basis for working in oil. "I tell people I'm a petroleum engineer and they just give me the stink eye," he said. "They think I'm evil. [They think] 'Oh, earth raper,' right?" Rather than debate them, he hoped that his work would change their impressions: "I don't vocally get up there and get in debates with people about [the oil and gas industry]. I just do the best that I can and let my work speak for itself."

Colton believed that truly irresponsible oil and gas companies were few, but his career satisfaction was still grounded in being able to choose clients and projects that aligned with his own senses of accountability. "I'm not conflicted about my own role in the industry," he said, "because I have the option to participate or not participate in that kind of [irresponsible] operation and I won't." To illustrate the point, he relayed a story of a Chinese company he witnessed operating in Kurdistan that brought in extra workers, assuming that a large percentage of them would experience death or injury on the job. "So when people aren't valued as your most important resource, I stay away from that operation, right?" Colton's language emphasized separation between his work and less scrupulous industry actors, saying, "I'm able to distance myself from that in that respect. . . . I've been able to decouple myself from that." In Colton's case, consulting allowed him to decouple from a larger industry and invest his time and energy into making his own projects as responsible as possible.

Lila, an energetic engineer, sustainability specialist, and entrepreneur, also chose consulting as a way to choose projects that aligned with her own accountabilities. She described herself as having "always wanted engineering to help people," which inspired her to pursue the most nontraditional engineering degree of all the engineers in this research—a degree in "ceramic engineering and society" that in her case included substantial elective coursework in anthropology and women's studies. Her first job after graduation, however, was developing new colors for automotive parts, which she dryly described as "not full of passion." To be closer to her family, she took a job at one of the largest engineering consulting firms in the world, where she found herself bored developing and designing the brick linings for furnaces. On a whim, she attended a lunch hosted by the firm's sustainable development group and gradually began moonlighting on projects for them. Eventually she took a leave of absence so that she could pursue a master's degree in international development and complete six months of fieldwork in a marginalized community affected by a major mine.

After finishing her degree, Lila went back with a "mission" to the sustainable development group at her previous firm to engage in community relations work. Yet she eventually left the firm entirely because she detected a lack of desire on the part of the managing directors to take community relations seriously. She then worked on two international projects for a small consulting firm before moving back into full-time corporate employment, persuaded by a charismatic new director of community relations for one of the world's largest and most controversial mining companies who wanted to "turn the company around." In a particularly painful example of the distributed agency that characterizes corporate forms, her director's boss soon asked her to cover up human rights abuses and began harassing her when she refused, prompting her to leave the company. "I knew [the company] was doing things wrong, in violation of human rights even, and I couldn't do it anymore. So I left them. I parted ways. Then I started working for myself." With the sting of her last experience still fresh, she recalled telling potential collaborators, "I'm done with mining. There's no ethics here. You can't have a bone of integrity. I'm getting out of mining."

Only after Lila met a visionary with a different business plan—one that gave communities part ownership of the mines—did she accept another mining consulting job. The project convinced her to reconsider her previous strong stance on not accepting mining and to take on clients she trusted and respected. She distinguished her clients from others in the industry by saying, "The clients that I tend to meet and work with are kind, ethical as you can be. . . . We don't print lies. Well, the ones I've worked with don't [lie]. People really want to do well, want their company to do well." During her career, Lila stepped in and out of corporate forms. In her view, consulting gave her greater opportunity to choose which "ethical" companies and personnel she would work with and on what kinds of projects she would dedicate her own efforts.

RISKS OF "RECOMMENDING" ACTION

While the consultants I met treasured the professional autonomy afforded by their ability to step out of corporate forms, many were simultaneously frustrated by the limitations that this liminal status presented for their ability to shape the behaviors of the corporate teams with which they worked. In other words, it was not always easy to detach their own personal and professional hopes, ambitions, and desires from the *projects* they worked on, even if they welcomed detachment from the *companies* themselves.

Jennifer worked for a major consulting firm that specialized in tailings dams, one of the most controversial and potentially catastrophic elements of mine infrastructure. When reflecting on the social responsibility dimensions of her career, she began by pointing to the structural position of her work: "From our position as a consultant, it's difficult to, but we can, influence change. But, you know, ultimately we're not the ones footing the bill for this project. So we can recommend lots of different things. But ultimately, we're not the ones that are building it, operating it, owning it. So we're in a difficult position when we see [that] it would be so much easier if they would just do what we recommend." Fabiana Li found similar concerns during her research on mining conflicts in Peru. Because consultants were limited to particular projects, they rarely got to see the "complete picture," in the words of one of the engineers she interviewed.[24] Echoing chapter 3, Li found that these professionals were asked to bring their expertise to bear on questions of how things should be done, not whether they should be done. One engineer explained: "As consultants, we limit ourselves to the question that is put to us."[25] Another of her interviewees was direct: "We are engineers, and we are an engineering firm. Our commitment is to the client, and we have to help clients carry out their project. We [cannot] get fundamentalist with environmental themes."[26]

Peter, an environmental engineering consultant, also experienced frustration when companies did not take his advice that they ostensibly hired him to provide. He began his career in the 1970s, highly skeptical of corporations. Describing himself as a "closet tree hugger," he vividly remembered coming of age as the "Cayuga River set Ohio on fire, and you'd see pictures of these four pipes discharging colored dye into the rivers, and it was just appalling." That environmental consciousness made him highly critical of mining.[27] Yet a chance encounter at the university where he was pursuing a PhD in engineering led him to a summer job on the Minnamax project analyzed in chapter 3. As an employee of a state agency, he helped set up the water quality monitoring studies with AMAX and enjoyed the work so much that he went on to accept a full-time job with the state and spent the next three decades working on mining projects. In our conversations, Peter positioned himself as an outsider to industry. He described common sentiments among environmental engineers while he was beginning his career by saying, "You just felt like you were basically doing battle with the evil industry. You know, you got to battle the evil industry and you don't want this huge mine on the border of the Boundary Waters Canoe Area polluting the water because all mines pollute." He remembered other mining areas that "literally looked like a moonscape . . . because it was so rocky, because all the vegetation had been killed."

From his position at the state agency, Peter set up some of the world's first studies on waste rock leaching. Even though the scientific literature at the time advised that tailings with less than 1 percent sulfur were safe, many problematic cases had emerged that were under the 1 percent threshold. Because of his research experiments with AMAX, Peter was able to demonstrate that the waste rock was contributing to release of acid and thus convince companies and industry regulators to change best practices for handling waste rock. He went on to establish more cooperative research projects between the state and other mining companies and their consultants, taking on the position of a sympathetic yet supportive industry outsider. What he found meaningful in his work was that he could "actually make a difference," while he thought his graduate work the lab "wasn't going to have much impact on my life or anyone else's life." He explained:

> We [at the state agency] were working on projects with the industry, and we were trying to solve problems. The mission of our group was to support the environmentally sound development of Minnesota's resources, and that includes mining. And so it was a very different mindset. I mean, as I said, I mounted up on my white horse and charged to do battle with the evil industry, and then all of a sudden, here I am saying, "Well this is a big change, you know? I'm just trying to solve their problems so they don't make a mess."

Peter eventually left the state agency, when he felt that it had come to lack "flexibility and commonsense" after being taken over by people he viewed as anti-mining advocates who were more concerned with avoiding potential lawsuits than actually promoting the responsible development of resources. He moved into consulting work, first at a large, multinational yet employee-owned consulting firm and then at smaller firms as he began transitioning into retirement. What he appreciated about his work as a consultant was the relative independence he enjoyed, saying that he could turn down projects that did not align with his own values: "I don't have to take anything." While he found it painful to observe mining projects "going down the toilet" from the position as an outsider, he said his reputation allowed him to choose his clients and avoid those that did not share his commitment to the environment.[28]

Indeed, the consultants I met who said they rarely experienced alienation from the implementation of their work were those who had a steady client base and could therefore be choosier about accepting contracts. Jen, for example, built up her own social and environmental consulting company and found success working in some of the world's most challenging social, political, and economic environments. What she loved about her work as a consultant was that it allowed to her ask the kind of big questions about mineral development that inspired her, such as, "How do you translate mines in poor countries or mines anywhere into poverty reduction?" She said, "I had this feeling that we knew how to make money, we know how to create wealth, but we didn't know how to reduce poverty." Even as an undergraduate mining engineering student, she saw how worksite engineering decisions that were presented to students as "technical" decisions informed by economic considerations were also inherently tied up in the wider social context of the communities close to the mine. She recalled being taught the benefits of block caving, an underground mining technique that involves undercutting the ore body so that it slowly collapses under its own weight. Within industry, the technique is referred to as the "underground version of open pit mining" and is praised for requiring fewer workers than other techniques. But seeing mining as an inherently sociotechnical endeavor, Jen immediately worried about what the decrease in employment would do for the mine's social contract with the people who lived and worked nearby: "If you took away one piece, if you took away the employment, how could [the mine] justify itself? And what was it actually delivering for a community when it was costing a community something, and costing a society something, and taking a finite resource away? What was it actually delivering back to the nation, back to the community?"

Jen's big questions about poverty led her to obtain a master's degree in development after finishing her undergraduate degree in mining engineering. She then took a job for a major mining multinational, but she left because she came to realize that the company's

> first mandate was always to be a mining company with their development contribution to communities and countries only a second order consideration. . . . So I thought I'd work for myself so that I could pick and choose, and pick the projects that were companies that I felt were ethical, or that were trying to do the right thing, or were trying to be brave and do something that was outside the box . . . I wanted to have the freedom to choose the projects that I would work with. I wanted to work with the ones that I truly believed had development potential and that were going to contribute positively.

Jen's emphasis of her own agency in leaving the corporate world in order to be able to choose projects and clients threaded through the narratives of the consultants I met. Her experiences highlight that this relative autonomy depended on having a steady enough client base that she could turn down projects in which she was uncertain about the clients' commitments. "If there's a company that doesn't appear to be at all interested in my recommendations and is just asking me to do something to get a report written to show that someone did something, I don't take the work in the first place," she said. Her favorite clients were those with whom she had developed mutual trust so they could work out problems together, "to be in that place where people are asking for your help rather than, sort of, 'We've got this piece of work that we think we ought to do. Can you please do it and here's this really defined scope?' I never get that." Jen wanted to participate in the scoping—or what engineering educators might call problem definition—rather than being handed a narrow task. She knew that it was in those initial stages where development concerns could either be made central to the question at hand or marginalized.

In narrating her motivations choosing projects and shaping them, Jen used the language of improving the industry and its contribution to development.[29] She described her ideal position as one in which she thought she could "actually help." She said, "I didn't want to just make someone look good because . . . they spend some money on a project. I didn't want to be a part of that. I wanted to be with people who were really trying to fight to make things better." This narrative of reform was prominent in how consultants wrestled with their participation in industries that they viewed as holding promise to improve people's lives, despite the prominent cases of companies failing to deliver on those promises.

"STEERING THE SHIP": NARRATIVES OF REFORM

If consultants desired professional autonomy and many had deep reservations about the industries in which they worked, why did they not leave industry completely? What were their attachments to these industries in general or to the companies they with contracted in particular? In this vein, the most common theme that emerged throughout my research was a narrative of reform that acknowledged but also critiqued the ethic of material provisioning analyzed in chapter 2.

Scott, a midcareer geological engineer, was skeptical of how profit motives shaped the practice of engineering, but he maintained his work with corporations through his position at an independent geotechnical consulting firm. Among my interlocutors, he was the most explicitly reflective in our conversations about when he was speaking on behalf of himself, his consulting firm, or the project as a whole: at multiple points in our interviews, he said *we* but then paused to clarify to which "we" he was referring—his consulting firm or the project team, which would include the company personnel. Scott said he appreciated the opportunity to develop relationships with and make changes inside of powerful companies, but he acknowledged that "as a consultant, your ability to change some of the big-picture issues is really limited." His main concern was that for-profit companies were "kind of doing the minimum, from what I can tell, the minimum they need to do to have the social license and the actual permits that they need to develop the project. So I get the sense that everyone is sort of representing themselves and just doing the minimum required to make the project move forward." In contrast, he described his colleagues from the firm as sharing a similar progressive political ethos.[30]

Yet Scott believed firms such as his played an essential role in ensuring that natural resource production was done responsibly. His specialty was roads, specifically, making recommendations to mining and oil and gas companies about where to place them to minimize financial costs of construction along with exposure to geotechnical, hydrotechnical, and geohazard risks. He saw roads as "intrinsically political," recognizing that his infrastructural work as always embedded in social structures and relationships of power.[31] For mining and petroleum projects, the recommendations that he and his team made about where to place access roads and where to cite major infrastructural elements such as tailings dams facilitated large-scale industrial development where there previously had been none—a conundrum that sparked lengthy soul-searching conversations among his coworkers at the office as well as at their field sites. He described his coworkers at the consulting firm as "people who are excited about the outdoors, like me," only to find themselves directly contributing to the transformation, if not destruction, of the environments they value.

> So then, we all arrived at [a field site] and we sit around and we say, "Well . . . but now, here I am, going to this pristine wilderness. I get to fly over in a helicopter, it's amazing. But we're here to . . . I mean, destroy it." We think that. Like, "You know, it looks so amazing now, but ten years down the road, this is all gonna be a tailings pond and you won't recognize it." And we're all sad about that. And so, yeah, we sit at these camps in the evenings and lament about the fact that this beautiful valley's gonna be destroyed. And then, we talk about, "Well, how do we justify doing this?"

Scott was not alone. Dan, a fellow geological engineer, had worked both as a consultant and as a full-time corporate employee. He said, "We all got into geology and geotech because we love the outdoors, but we discovered that the resource industry is the only place to work. We rationalize it by saying that we're all users of resources and therefore our responsibility is to do that development responsibly." Scott justified his career decision through the metaphor of "steering the ship," the ship being a minerals industry that increasing levels of consumption around the world made necessary:

> It comes back to the argument . . . I flew here in an airplane, I drove my car across town. I have a standard of living that I've grown used to that I'm not gonna walk away from. And so, if I'm gonna have those things, then I need to be part of the . . . we say, "steering the ship." We got this big ship. And either you can be the Greenpeace who jumps in front of the ship and tries to stop it—and I'm fully supportive of those people—but I could choose to be that person who is jumping up and down in front of the giant ship and gets pushed out of the way, or I could help steer the ship. So we say, "Well, all of us at [our firm], we're helping to steer the ship. And maybe we're steering the ship off the edge of a cliff, but we're steering the ship." And that's how we justify it, rightly or wrongly.

In this statement, Scott acknowledged the dominant ethic of material provisioning, pointing to his own consumptive practices that made mineral development necessary, but he simultaneously recognized it as a particular framing of accountability rather than as simply the "truth" about how the world was. He also subtly critiqued this ethic by opening up the ultimate "good" it proposes to question, jokingly saying that they might actually be steering the shift "off the edge of a cliff."

To help steer the ship, to make the industry as responsible as he could from his position, Scott paid attention to "really consequential stuff" that appeared deceptively "mundane." One of his projects at the time of our interview, for example, was assisting a South American mining company in decommissioning the access roads and drill platforms it had been using to explore a mine expansion that never materialized. The original inhabitants, many of them campesinos, had returned to the exploration sites to find new flat pads, in the midst of an otherwise mountainous Andean landscape. They took advantage of the pads and constructed homes and small businesses, but those pads were flanked by steep and potentially unstable cut slopes designed for short-term use by the mine's subcontractors wearing personal protective equipment, not for long-term habitation by people untrained in landslide management. When Scott and his team from the firm arrived on site, they saw that the risks to the returning residents were not included in the original proposed scope of work. He and his coworkers convinced the people in charge of the project to broaden that scope to include a risk assessment to investigate the possibility for people to live long term in those areas. When we spoke, his team was studying the dozens of drill paths and roads to prioritize those that posed the greatest risks, so that they could recommend plans and best practices for how to close those areas or lower the risk to a tolerable level for long-term human habitation.

Scott, however, was unsatisfied with just trying to make corporations more responsible and established professional spaces that were not dictated by profit demands and allowed them to serve people who had been underserved by engineering. The part of his job he loved the most was helping found a division of his firm that would engage in pro bono work for the communities that needed geotechnical expertise but could not afford to hire their own experts. "We have to have enough corporate clients so we can pay our bills," Scott explained, "but we have more than enough business to reinvest our earnings back into communities." Because the firm was owned by its employees, they had the power to create the division and direct a substantial percentage of their profits to support its work. For Scott, the "person" of the consulting firm he enacted was still constituted by multiple parts, like the corporations he critiqued, but in the consulting firm he felt that those parts nonetheless shared many political and social commitments.

Like Scott, Gary craved professional spaces that would directly serve marginalized communities. After becoming disenchanted with the corporate world and leaving his consulting job, he formed his own nonprofit that would put geotechnical knowledge at the service of communities that could not normally afford it. Motivated by big questions about economic development, social justice, and environmental sustainability, Gary dreamed of finding a way for poor communities to process tailings—heaps of accumulated mine waste that still include trace "uneconomic" minerals left behind after the main mineral has been extracted—for smaller amounts of profit that could nonetheless make a big difference for the poor.[32] "There are resources in those materials, but they're not enough to trigger the greed in a mining company to do something with it," he explained wryly.

While Scott and Gary both created professional spaces that were more distant from the corporations they contracted with, other engineers took an opposite career trajectory and left consulting to enter full-time corporate work specifically to influence change from inside of powerful companies. Benjamin spent his first years as a consulting environmental engineer helping mining companies demonstrate regulatory compliance. The experience helped him develop expertise in geology, minerology, sulfides, acid mine drainage, water quality assessment, impact prediction, and closure plans. A participatory environmental monitoring assignment in the Global South impressed on him the importance of doing technical work that mattered to the people who bore the greatest burdens of industrial development: "Suddenly you're working to collect information, make opinions that people really care about. It's not just making opinions for a regulator to sign off on to put the book on the shelf to show that the process has been done so that then you can do whatever activity you want to do. But rather, people really care about the results because they feel like it impacts them directly, right?"

The experience compelled to work with "clients that wanted to do technical work or environmental work but with community interests and community involvement." He said he eventually took a job with one of the world's largest oil companies in order to make a bigger impact:

> Then I did think that one of the gaps in my experience was working for a company, right? Because you're always working in consulting. . . . You're collecting information. You're synthesizing, you're writing, but you're not actually working at the very up-front end of the companies that create the problems in the first place, or however you want to couch it, right? So it seemed like a good opportunity to do similar work but internal to a company and hopefully make a difference.

Similarly, Paul was a senior environmental engineer who moved from a successful consulting career to one of the world's largest mining multinationals because he could not pass up the chance to spearhead a major cleanup effort. In narrating his career, he signaled his own critical assessment of the company and industry as a whole. He described himself being very resistant to a friend's suggestion that he interview with the mining company, recalling that he said to him, "There is no way in the world I will go to work for a mining company. I will damage my own personal reputation, and it's just not going to happen." He said he "reluctantly" agreed to job interview but became convinced of the magnitude of the project they wanted him to lead—a cleanup that would go on to become the largest privately funded cleanup in US history—and company's serious commitment to doing it right. "So I ultimately sort of swallowed my pride and said, 'Yes, I'll come to work for you,'" he said. "And sure enough, the company put its money where its mouth was." Even as a junior engineer, Paul felt that his colleagues listened to him and respected him, so much so that they flew him to the corporate headquarters to talk to the board about the cleanup program. "To me, that was something I had never experienced before," he said. "And what impressed me the most was that the board was knowledgeable and interested and asked good questions and pushed me in certain areas, but it was obvious that they took this all very seriously." Many engineers and applied scientists were like Paul in portraying themselves as initially skeptical, as described in chapter 4, perhaps because doing so both emphasized their own agencies of critically assessing companies' performance and leant further credibility to the responsible image of the companies they sought to portray.

By the time the cleanup was finished, Paul said, their team had moved "more hazardous waste material than the entire EPA Superfund program across the country." He went on to management and executive roles in the company, helping establish a climate change policy that was the first of its kind for a mining company and to integrate environmental sustainability and social responsibility into operations, including the decision-making process for new projects. His corporate career allowed him to become a powerful agent of change inside of industry, though he experienced unease when visiting some of the company's less respectable operations that were just giving social responsibility "lip service," he said. "They weren't really doing the work." His frustration points to the tensions that emerge in corporate forms that are constituted by multiple agencies—the "laggers" reflected poorly on Paul's performance because the company was collectively constituted, even though it was geographically dispersed across the globe. To try to align the company's agencies with what he viewed as his and coworkers' more responsible practices, he tried to mentor the laggers to embody the company's overall social responsibility framework. Paul said that, while it was demanding work, he appreciated the opportunity to transform the industry practices he viewed as the most troubling.

"WE'RE BASICALLY HERE TO SERVE": NEW SOURCES OF LEGITIMACY

Stein argues that corporate clients enroll consultants to shore up their own authority to fight in-house battles, positioning them as "outside" experts who endorse or even propose the client's own desires for change.[33] In this section, I show that corporate actors also enroll consultants as key sources of legitimacy for *external* audiences as they manage their accountability to members of the public. Yet while company personnel and public relations offices tout the detachment of consultants from corporate forms as an "independent" stamp of approval for their projects, the structural conditions under which consultants work engender dependence on those corporations for continued contracts, constraining that very independence. In other words, corporate personnel can claim the objectivity and independence of consultants as a way to shore up public trust in their own projects at the same time as those consultants' activities are constrained by their dependence on those same corporate forms.

In my research, tracing out the contribution of consultants to the everyday activities of corporations and their representational practices was difficult given how deeply they were embedded in most industry activities. Sometimes the knowledge that engineering consultants produced was enveloped into corporate self-representational practices, such as when corporate personnel shared the results of environmental studies in public meetings while attributing them to the *we* of the corporation without signaling that they were completed by consultants hired by them. At other times, corporate personnel made visible the consulting firm origin of those studies, perhaps because it added a veneer of objectivity—assertions that a mine would not compromise a groundwater table seem more believable when coming from an "independent" consultant from the same area as the groundwater table than from a distant corporate entity proposing the mine.

During my research, the most discussed US mining controversy was the proposal by PolyMet Mining Corporation to build a large open-pit metal mine in northern Minnesota, in close proximity to the Boundary Waters Canoe Area. A 2012 PolyMet news release titled "PolyMet Strengthens Permitting Expertise—Groundwater Monitoring Requirements Satisfied" highlighted the role played by consultants in the project:

> Foth is a recognized leader in environmental review and permitting with a nationwide reputation for successful permitting of nonferrous mining projects in the upper Midwest. In particular, Foth was the lead consultant that secured the environmental permits for the Kennecott Eagle nickel and copper mine in Michigan and the Flambeau Copper Mine in Wisconsin. "I am very

excited to have Foth join our permitting team," said Jon Cherry, president and CEO of PolyMet. "Foth's key role is to provide strategic advice related to securing the permits necessary to construct and operate the NorthMet Project in an environmentally sound manner, as well as to ensure appropriate quality of permit submittals."[34]

Consulting engineers both from Foth and from Barr Engineering took on public-facing roles as the debate unfolded. The tailings dam became one of the most controversial components of the proposed mine, in the wake of devastating dam failures in Canada and Brazil. Barr Engineering was the lead firm in charge of the PolyMet dam project. Scott Grosser, a geotechnical engineer for Barr, appeared in multiple news articles to vouch for the safety of the proposed dam, including a 2017 article in the *Duluth Tribune* in which a PolyMet official specifically highlights the inclusion of consultants as improving the safety of the project:

> *Both internal and outside engineers*, including consultants for the DNR [Minnesota Department of Natural Resources], all have suggested modifications to the design that will make it stronger and better, said Brad Moore, PolyMet's executive vice president of environmental and governmental affairs. That includes dropping a plan to use concrete pumped underground to shore-up the base of the dam and instead use rock on the outside of the dam to make it easier to monitor how the buttressing is working and allow more to be added if needed. "The tailings basin design was one of the most-studied aspects of our project," Moore said.[35]

The "Tailings Basin Stability and Environmental Protections" information sheet that PolyMet produced and made available on its website also specifically calls out the role of Barr and other consultants:

> Geotechnical experts independently performed numerous geotechnical evaluations and concluded it is feasible to add our tailings to the existing facility. These experts will assist in the specification of future engineering requirements prior to and after production begins. . . . Minnesota-based Barr Engineering, our engineering firm for the tailings basin, has been designing tailings basins for mines in northern Minnesota since the 1960s and now works on tailings basin projects around the world.[36]

Yet other information sheets took on a third-person perspective that did not make visible the studies contributing to its assertions, such as about water quality or financial assurance. The same third-person perspective runs throughout the environmental impact statement, which itself was prepared by the consulting firm Environmental Resources Management at the request of the lead government agencies on the project. The executive summary—the section that was most accessible to readers out of an environmental review totaling a staggering twenty-two hundred pages[37]—asserts conclusions without citing the original sources at all. For example, it argues, "With the proposed engineering controls, the water quality model predicts that the NorthMet Project Proposed Action would not cause any significant water quality impacts."

A senior civil engineering consultant who worked for one of the firms hired by PolyMet was critical but pragmatic about the role his firm played in industry. Stuart's specialization in groundwater and soil contamination opened up ample opportunities for him to critique companies, which he summed up by stating, "A client will push you right to the ethical edge of your credibility." His long-term perspective on working with a variety of clients and holding executive positions at the firm provided him insight on the asymmetries between clients and consultants. Stuart believed that major companies hired consultants because "they want the local relationship, they want the local credibility." He traced this strategy all the way back to AMAX's Minnamax mining exploration project in Minnesota in the 1970s, where he saw the New York–based company shore up its credibility by emphasizing the work done by his consultant group based in Minneapolis. He speculated that skeptics of the project would be more likely to trust the Minnesota firm than the New York City conglomerate, as would the state regulators who would eventually be tasked with approving permits. Forty years later PolyMet engaged in similar representational techniques: the same 2012 news release that praised the consultants hired by PolyMet played up Barr's "local" identity, referring to the firm as "Minnesota-based Barr Engineering" and its "experience in Iron Range mining projects."[38]

Stuart criticized the structural position of consultants in relation to powerful mining companies like PolyMet, stating, "We're basically here to serve. They tell us what they need, and we develop a work plan for doing that. They may ask us what we think once in a while . . . [but] they're the client and we're the consultant and they're running the project." Stuart's critique resonates with scholars who examine the limitations and opportunities of engineers being "designed to serve."[39] Such an orientation has positioned engineers as solving problems defined by others rather than defining problems themselves.[40] This is problematic because the problem-definition space is where wider concerns about responsibility, accountability, or justice could be defined into or out of the equation, as consultants like Jen recognized. Being excluded from problem definition undercuts the ability of ostensibly "independent" consultants to practice engineering in a way that resonates with their senses of social and environmental responsibility. The permeability of the corporation—and its enactment by consultants who were structurally positioned to serve industry—served a strategic purpose in accessing new relationships and new sources of credibility for the corporation's activities, even as corporate personnel maintained the prerogative to do as they wished with the consultants' recommendations.

CONCLUSION

The engineers I met who were most critical of corporations were those who worked as consultants, from Gary's condemnation of corporate America to a seasoned petroleum engineer who believed that all natural resource conflicts could be reduced to "lawyers, guns, and money" and that young engineers should be taught that they are nothing more than a "tool of US foreign policy" that exists solely to further the interests of the shareholders of major transnational corporations. Consulting seemed to offer more attributes that scholars would associate with professional autonomy. Most significant, these engineers and applied scientists had more opportunities to choose to work for clients and on projects that aligned the most closely with their public, professional, and personal accountabilities, from Colton's commitment to create operations that would not offend the environmental sensibilities of outdoor enthusiasts like himself to Gary's ambitious goal of restructuring natural resource production to benefit poor communities. This can be interpreted, I suggest, as a strategy to navigate the challenges of organizational work, specifically their being held accountable for the multiple enactments of an extended corporate "person."

The emphasis these consultants placed on projects that dovetailed with their public accountabilities suggests that "meaningful" work was that in which public accountability was integrated into the engineering itself. This finding builds on Peter F. Meiksins and James M. Watson's conclusion that "engineers, while not entirely comfortable with organizational constraints, view them as inevitable and are willing to accept them in exchange for interesting work."[41] This chapter shows that, at least for one group of engineers, "interesting" work was that which spoke to their sense of accountability outside of a narrow technical realm. Engineering professor James Trevelyan reached a similar conclusion in a large study that comprised over three hundred interviews with practicing engineers, survey data from nearly four hundred engineers, and multiple years of participant observations of Australasian engineers. He found that "more experienced engineers, those who had stuck with it for a decade or more, had mostly realized that the real intellectual challenges in engineering involve people and technical issues simultaneously. Most had found working with these challenges far more satisfying than remaining entirely in the technical domain of objects."[42]

The chapter also suggests that the boom of consulting opportunities may have changed the trade-off originally proposed by Meiksins and Watson, as full-time corporate employment no longer seems as inevitable as it once was, given the swelling ranks of consultants. But while consulting opened up many opportunities for engineers to work for corporate clients they respected and on projects they admired, they also expressed frustration in not controlling how their work was eventually implemented by those clients. Consulting thus seems to introduce a particular kind of alienation for engineers beyond what engineering studies scholars have influentially theorized as estrangement from craftwork or the "professional-bureaucratic dilemma."[43] Many consultants experienced alienation from the implementation of their work, from how their studies were represented in an environmental impact statement to how a tailings dam was eventually constructed. Only a few were willing to seek full-time corporate employment to try to ameliorate such alienation. It was far more common for consultants to try to select corporate clients as carefully as possible.

The political economy of consulting raises serious questions about the accountability of corporate forms. Writing about environmental consultants who work on mining projects in Guatemala, sociologist Michael L. Dougherty describes the double bind they face: "They want to exercise their technical expertise and advocate for stringent and effective environmental management, but they recognize that their solvency depends on shilling for miners."[44] While the consultants I met all positioned themselves as scrupulously evaluating potential clients to ensure an ethical fit, only the most successful were able to build up enough relationships and projects that they could afford to say no to those that offended their sense of professional, personal, and public accountability. The fate of those ethically suspect clients and projects points to another dispersal of accountability: even if one consultant refuses the work, another one can pick it up.

The ambiguous fate of the consultants' work provides a vivid example of how these more "autonomous" professionals are nonetheless enfolded into corporate practices of accountability that extend the authority of corporations themselves: the consultants' studies, plans, and reputations were put to work in the service of their corporate clients. Corporate personnel highlighted the independence of consultants as a way to bolster support for their projects, even as consultants remained dependent on them for continued work. This focus on the liminality of consultants underscores that the partible and permeable nature of corporations contributes to making this form so slick and powerful.

For their part, most of the engineers found some peace considering—or perhaps consigning—themselves to be reformers, not revolutionaries. Accepting the inevitability of working for corporate forms in some way, they sought work that allowed them to practice engineering in ways that resonated with their multiple accountabilities. Their strategies seemed to echo what Harvey and Knox identify as the relentless pragmatism of engineers.[45] Chapter 6 explores the politics of this pragmatism in detail, focusing in particular on how engineers in Colorado used their professional practices to attempt to quell the growing firestorm surrounding fracking in the state. It shows how attempts to cultivate accountability between engineers and the broader publics affected by their decisions rested on notions of reconciliation that marginalized criticisms that could not be reconciled with continued oil and gas production itself.

6 ENGINEERING PRAGMATISM

Facing the growing firestorm surrounding fracking in Colorado, petroleum engineer Aaron made listening the platform for his activities as he set out to change how his industry engaged the public. In speaking to students in my Corporate Social Responsibility course, he argued for the importance of listening by saying, "When you think you're going to change someone's mind by 'educating' them, you are assuming that the problem is in *their* understanding and comfort rather than *your* actions." Aaron was criticizing a common refrain among industry practitioners as well as engineering faculty and students that it was possible to change people's minds through "educating" them. His point resonates with a long line of science and technology studies scholarship that critiques how such an orientation positions the public as deficient while reinforcing the authority and expertise of scientists and engineers.

Aaron's statement was uniquely forward looking in the midst of growing national debates about fracking. Oil and gas associations and companies had otherwise embarked on what they called "educational" campaigns to use scientific "facts" to convince the public of the safety of their technologies and practices. For example, Energy in Depth, a group sponsored by the Independent Petroleum Association of America, created a web page that provided counterevidence for the claims vividly presented in *Gasland*, the 2010 documentary that put fracking on the national political agenda and inspired waves of anti-fracking movements. The American Natural Gas Alliance sponsored the filming of a rebuttal documentary they unironically called *Truthland*. In Colorado, the largest oil and gas operators joined forces to create Coloradans for Responsible Energy Development (with the "credible" acronym of CRED), a group that served as a clearinghouse to "provide information to the public about the economic and environmental benefits of safe and responsible oil and natural gas development."[1] Their news releases and website underscored the validity of their perspective by appealing to scientific evidence, encapsulated in such titles as "Scientists Agree: Fracking Doesn't Harm Our Water."[2] This ethos also permeated my own institution, the Colorado School of Mines, where well-intentioned engineering students and professors adamantly believed that anti-fracking activists would cease their criticism if they just "understood the facts" about the safety of fracking.

Aaron signaled that his commitment to listening to critics raised controversy among his coworkers, saying, "It was as if by *listening* to 'those people' we were sympathizing with them" (see chapter 1). Years later, the tenor among industry personnel had changed at least partially, especially in corporate outreach materials and official presentations. Professed commitments to listening to stakeholders pervaded wider corporate social responsibility (CSR) discourses, along with admission of past mistakes and promises to improve.[3] Engineers were quick to point out that good listening could not make up for irresponsible engineering. An engineer and executive summarized this philosophy by stating, "You can't talk yourself out of a problem that you engineered yourself into." Aaron shared a similar opinion, emphatically stating that "listening is not a surrogate for quality operations."

For Aaron and other engineers I met, listening was important because it was how they could glean information to help them make engineering accountable to the public. They desired to know more about the local context of oil and gas activity so that they could attempt to reconcile industrial activities with local concerns—what some of them termed generating "compatibility" between industry and communities. In this chapter, I argue that although this everyday practice of accountability financially benefited some parts of the public while allowing companies to maintain or expand their reach, its focus on "actionable feedback" foreclosed broader questions about industrial development. Whereas engineers viewed these as win-win scenarios that benefited companies and publics, they sidelined the concerns of people who wished for no industrial development at all.

These practices of accountability configure engineers' agency in particular ways. Chiefly, I show that they reflect and foster a deep-seated engineering pragmatism.[4] In its academic sense, pragmatism refers to a philosophical tradition emphasizing usefulness, workability, and practicality.[5] That tradition echoes in humorous jokes about engineers who are so committed to solving problems that they endanger their own well-being, and it animates popular distinctions between engineers as driven by application and scientists as driven by curiosity or truth.[6] Though my interlocutors also shared in this general appreciation of practicality, in this chapter I identify a more specific sense of pragmatism animating engineers' work in corporate settings. The engineers faced a dilemma that has been left unresolved after nearly a century of debate about the engineering codes of ethics analyzed in chapter 2: their professional norms held them accountable to both their corporate employers or clients and the safety, health, and welfare of the public, providing no guidance about what to do when those two accountabilities conflict. In my research in particular, engineers found themselves simultaneously facing public calls to change industry practice and pressure to keep the corporate forms employing them financially solvent. In an attempt to be accountable to both demands, they tried to make their engineering decision making both profitable for their companies and responsive to public demands. It is this particular effort at compromise that I signal when using the term *pragmatic* to describe engineers and their practices.[7]

This pragmatism is worth exploring at length because it can exist in tension with more radical forms of action and critique, especially as they relate to social, environmental, and technological upheavals. Indeed, Carl Mitcham proposes viewing the Anthropocene as the "Engineering Epoch," given the significant role engineers have played in creating and maintaining the infrastructures and forms of knowledge that created this era.[8] While there is much to admire in engineers' pride in being problem solvers, this intertwined sense of self and profession can also privilege maintenance of the status quo rather than fundamentally questioning it. As a case in point, the engineers I met almost all envisioned a more responsible—and thus more robust—form of capitalism rather than the "potential of capitalism's undoing."[9] The practicing engineers I met and most of the engineering students I taught were willing to entertain the kind of sustained critique that drives social scientists but grew frustrated and unsatisfied when that critique could not be easily translated into some sort of productive intervention or improvement. They wanted to know what to *do* with the critique, how to act on it to address whatever problems it revealed. While this deep desire to help may be admirable in its aim to improve people's lives, such projects of "improvement" are never politically neutral.

In this chapter, I focus primarily on the fracking firestorm in Colorado to show how this engineering pragmatism can end up sidelining concerns that call into question the continued need for natural resource production itself. I draw attention to the politics of listening, specifically, how some concerns exceed institutional constraints on what content could become "actionable feedback." This analysis draws attention to the limitations of engineering agencies and therefore of their accountabilities, building on the theorization of the distributed nature of the agencies that constitute corporate forms and extending the concern with detachment detailed in previous chapters.

LISTENING AND ENGINEERS' SOCIAL RESPONSIBILITIES

In the most comprehensive study of engineers and CSR, Gwen Ottinger shows that community engagement tools ultimately reinforce the scientific authority and expertise of petrochemical companies and their engineers working in Louisiana. The vibrant community-based environmental justice group she studied dropped its lawsuit and participatory air monitoring programs in the face of corporate commitments to "open dialogue" and the surrounding community's new feeling that they were listened to and respected by corporate actors. Ottinger argues that, in this context, appeals to communication reinforced a problematic and pervasive technical/social dualism by admitting mistakes in the "social" domain of communication, all while shoring up the company's own expertise and authority in the "technical" domain of operations and environmental monitoring. She writes: "Community grievances [were] framed as social issues, requiring thoughtful attention from managers, but no rethinking of technical practice. By taking on social responsibilities as part of their core business values, then, petrochemical companies created a space for plant managers to admit serious faults in their interactions with residents, and thereby resolve community conflicts, without jeopardizing their technical authority."[10]

The engineers I met sometimes similarly took shelter within such a technical/social dualism.[11] This dualism is predicated on depoliticization, or the belief that "technical" concerns can be purified out of their social, political, and economic context.[12] But those I met also sought to engage in professional practices that would be directly accountable to various publics. By pragmatically seeking to adapt the process of resource production to be responsive to the concerns of the people who would be most affected by it, they enacted a vision of accountability that was more sociotechnical in nature.

Their practices formed and were informed by shifts from what scholars call "old" to "new" CSR. New CSR activities change core business practices that create harm, whereas old CSR activities do not.[13] Philanthropy is the most prominent example of old CSR, as it can provide feel-good images while coexisting with the continued production of harms—picture Massey Energy CEO Don Blankenship handing out Christmas presents to poor children in West Virginia while mandating unconscionable safety cuts that would claim twenty-nine lives in the 2010 Upper Big Branch mine disaster.[14] In their ideal form, activities in the "new" model of CSR seek to address those harms directly by changing core business practices, ideally to internalize what might otherwise be externalities. In the mining and oil and gas industries, these core activities directly involve engineering, meaning that new CSR involved changing the dominant expectations and practices of engineers. For my interlocutors most committed to this version of CSR, listening was the mechanism through which they would integrate community concerns into engineering practice.

Engineering educators also point to the transformative potential of listening. Gary Downey advocates for positioning engineers as "both problem solvers and problem definers who listen."[15] Doing so also required him as a scholar to listen to the ongoing struggles of engineering educators themselves, to position his interventions in a way that they could be taken up as critical participation rather than dismissed as external critique.[16] Building on these perspectives, Juan C. Lucena, Jen Schneider, and Jon A. Leydens emphasize listening in their efforts to harness engineering to promote sustainable community development and social justice. Noting that undergraduate engineering training in "communication" privileges speaking over listening, they offer a theory of contextual listening distinguished from what they call basic listening. Whereas basic listening can be conceptualized as information exchange through output (speaking information) and input (receiving information), contextual listening involves situating such exchanged information within the broader historical and structural factors that make it meaningful: "Information such as cost, weight, technical specs, desirable functions, and timeline acquires meaning *only* when the context of the person(s) making the requirements (their history, political agendas, desires, forms of knowledge, etc.) is fully understood."[17] This kind of listening invites engineers to put themselves in someone else's shoes to understand not just what that other person is saying but the broader context giving shape to those statements.

Some corporate actors would likely look positively at the concept of contextual listening for providing richer information either to help them truly address people's concerns, in its most altruistic formulation, or to outmaneuver their opponents and co-opt their critics, in a more mercenary formulation. This underscores the fact that listening can be put toward different political ends, from neutralizing critique of corporations in Ottinger's work to promoting social justice for communities underserved by engineering in Lucena, Schneider, and Leyden's work. This raises the question of *for what ends* listening is being put into service.

The corporate context of many engineers' work further raises the question of scale. The community development or assistive technology projects analyzed by Lucena, Schneider, and Leydens, for example, are mostly predicated on small projects that allow engineers to communicate directly with the intended beneficiaries of their work. In contrast, the engineers I met who worked in a corporate context, whether as full-time employees or as consultants, found themselves facing significant institutional barriers to direct engagement with the people affected by their work. Not only did they work for large corporations with divisions of labor that tasked some employees with outward-facing roles while tying others to their desks, but the number of potential "stakeholders" for their projects was enormous, especially for the engineers facing the fracking controversies in the booming Denver metro.

ENGINEERING THE WIN-WIN

When giving examples of socially responsible engineering, almost all of my interlocutors pointed to projects in which they designed solutions that would create financial prosperity for companies and communities while minimizing the risks of industrial development—what they viewed to be a win-win but what social scientists would likely consider a harmony ideology (chapter 2). The question of what constitutes a "win" clearly depends on who is doing the defining. Aaron was careful to point out that simply mitigating risks and maintaining the same quality of environmental and social health was not a benefit to communities but an absolute minimum requirement to repair the disruptions they suffered.[18] The main benefits that he and other industry personnel saw for communities were economic gains in the form of taxes, royalties, and jobs. Some locals embraced such economic development, but others raised questions about the "logics of equivalence" that justified potential environmental and social harms by appealing to economic gain.[19] After all, some residents valued other "goods" than those proposed by oil and gas representatives: quiet evenings without the background noise of fracking operations, roads free of oilfield truck traffic, or minimal risk levels for air, water, or soil pollution.

In one sense, the projects of mutual benefit I analyze in this section were admirable in adapting technical practices to account for social and environmental contexts, unlike other practices of accountability that would cordon off engineering from its social context. Yet given the engineers' institutional locations as corporate employees responsible for producing profit, their practices of accountability were pragmatically aimed at continuing or expanding natural resource production. Like other harmony ideologies, the "win-win" marginalized more radical questions about resource production, conservation, and use in the problems that the engineers were attempting to solve through technoscientific creativity.[20] Here I group these practices of accountability into two broad categories: shared infrastructure and design for community acceptance.

Shared Infrastructure

My interlocutors referred to shared infrastructure as infrastructural projects that served necessary functions for both industry and nearby communities.[21] Juanita, a senior petroleum engineer who had worked her way into executive positions in safety and sustainability in oil and gas after beginning her career in mining, made a strong case for shared infrastructure and highlighted the more dialogical approach to community relations it required. Truly shared infrastructure, she said, required having conversations with local governments, planning commissions, and citizens to "find synergies." She continued:

> And it is a one plus one equals three if you do it right. In other words, if you're gonna build an airstrip, for God's sake, build it where it meets the longer-term needs of the community and spend an incremental amount of money to create something that's kind of fit for purpose for both uses. We [in the oil and gas industry] are all famous for building our own airstrips without having that [shared purpose]. The same thing with water systems, the same thing with power, increasingly with the infrastructure around telecommunications.

Other interviewees shared her enthusiasm for the transformative potential of shared infrastructure, especially surrounding increased access to the wireless communication and the internet in addition to rail, port, and energy access.

Though Juanita was referring specifically to development work in the Global South, she also saw opportunities in the United States, where cities needed to upgrade their wastewater treatment infrastructure and oil and gas companies needed to engage in water treatment. "There could be some synergy around waste treatment, waste-water treatment," she said and then joked, "that would be a whole hell of a lot better than taking clean water and pumping it down a hole and getting dirty water back and then pumping it in underground injection wells." Juanita contrasted the value generated by shared infrastructure with spending money on popular

CSR projects that were discrete from the company's core competencies and activities. She viewed the shared infrastructure projects as the "cheapest social investment you can do versus doing whatever the hell you want and then you sprinkle a little money around for a few schoolhouses or baseball fields." Her critique of "old" CSR echoed how an anthropologist, who had dedicated his career shaping the field of social performance in mining, made fun of old CSR in a campus lecture at Mines by saying, "Communities threw rocks over the fence at industry, and the industry threw back schools and hospitals."

Engineers who worked in mining were also quick to point to shared infrastructure as an example of social responsibility and supported their observations by drawing on a small gray literature including contributions from the World Bank and the influential development economist Paul Collier. The larger scale, longer duration, and greater capital investment associated with mining activity may make this a more common practice in that industry than in oil and gas. When I asked Jennifer, a geological engineer, about a good example of CSR, she immediately referred me to the wastewater treatment plant that Freeport-McMoRan constructed in southern Peru. The company was seeking to expand its large, open-pit Cerro Verde copper mine outside of Arequipa, Peru's second largest city, but could not do so without massively increasing the water it used. The region was already arid, and company personnel worried that creating a dam on the primary river would spark controversy. Jennifer explained that consulting engineers recognized that the city of Arequipa itself, home to about one million people, did not have a wastewater treatment plant, meaning that raw sewage was disposed directly into the river. She said that they then worked with the company to "come up with a solution where we designed a sewage collection system where we collect the raw sewage and take it to a wastewater treatment plant and treat the water. And then the treated effluent, some of it goes to the mine, and then the rest of it goes back into the river as cleaned—not drinking quality water but dischargeable quality—water."[22] The $500 million plant—part of a $5 billion overall mine expansion—was completed in 2015 and received international accolades for improving the health of Arequipa's residents, including a 2016 US Secretary of State Award for Corporate Excellence and recognition by the International Council on Mining and Metals and the Canadian International Resources and Development Institute as an exemplar in social and environmental responsibility. Yet Oxfam discovered that Freeport owed $250 million in unpaid taxes between 2006 and 2009, of which $140 million was owed to the local government, raising doubts about the company's actual commitment to public accountability.[23]

The increasingly popular notion of local procurement evinces a logic similar to that of shared infrastructure, though the "infrastructure" would be supply chains instead of physical installations. Local procurement involves companies purchasing goods and services from local businesses rather than from large companies owned by national or foreign elites. The goal is to direct as much of the company's financial investment as possible into the hands of local people, though such projects require significant training and face substantial bureaucratic hurdles in aligning local practices with industry norms and national laws.[24] As of 2020 local procurement was considered best practice for major extractive projects, with groups such as Engineers Without Borders–Canada providing analysis of how to make it benefit communities in a responsible, long-term manner. While some intended beneficiaries of these programs welcome them as a connection to global flows of capital, social scientists caution that such programs also promote an entrepreneurial ethos at the expense of other forms of claim making on states and corporations.[25]

Shared infrastructure helps address criticism from communities of the injustice of major infrastructural projects privileging industry rather than communities, as memorably captured in an ethnography of a village of Peruvian campesinos who lacked access to electricity while living underneath towering high-voltage electric lines that connected a foreign-owned mine with the national grid.[26] Yet the scale of these decisions tends to involve high-level corporate personnel and their government counterparts. It is near the executive level that major decisions about infrastructure are made inside of corporations, and the "listening" that seemed to inform them was directed at government officials. Juanita was careful to state that the synergies she praised could not emerge from engineers identifying a need and then "educating" communities about how to fix it but had to be based in a "real engagement" that exceeded instrumentalist desires to gain permit approval. But for her, the people she suggested that companies needed to listen to were government officials, who do not always represent the concerns of the full range of their constituents.

Design for Community Acceptance

A second area of socially accountable engineering encompasses projects I group as "design for community acceptance." Unlike shared infrastructure, the material artifacts and processes being designed are not for joint use by companies and publics. Rather, engineers factor community concerns and desires into their design of the material artifacts and processes to be used by companies themselves.

Marie was a petroleum engineer who had spent most of her career working in completions, referring to the phase of oil and gas development after drilling in which the well is brought into production, including through hydraulic fracturing. She described herself as a passionate advocate for the oil and gas industry and as proud to work for a company with a reputation for being progressive about securing the social license to operate. She spoke at length about how she had tried to integrate public accountability into her work. "I don't want to do a Band-Aid fix, just putting up walls and hay bales," she said, referring to common practices of mitigating the noise and visual disturbances of active wellpads. "I was a big pusher of, 'Let's reengineer the equipment because this will fix the problem for the next ten years, not just now.'" Marie's description fits within the broader discursive shift from old to new CSR, from trying to hide noise and visual disturbance to designing them out from the start. In these practices of accountability, engineers' understanding of community acceptance figures into engineering decision making itself.

Marie was also proud of being what she called the "main engineer" who created a "stimulation center" that reduced and spatially concentrated the overall footprint of the fracking process. Horizontal drilling allows operators to place multiple wells on one larger pad rather than spacing out single vertical or directional drills on multiple pads dispersed throughout farms, ranches, and communities (figure 6.1). While this consolidation may reduce

Figure 6.1
Hydraulic fracturing job in process in the Bakken field in North Dakota. Photo courtesy Joshua Doubek via the Wikimedia Commons: https://commons.wikimedia.org/wiki/File:Frac_job_in_process.JPG.

the number of people impacted by oil and gas production, it significantly intensifies the burdens faced by the people who live in close proximity to the enlarged sites.[27] Industry data for Greeley, Colorado, estimates that by 2018 multiwell pads were "commonly about 4 acres in size, holding 24 horizontal wells with associated equipment. According to industry estimates, well pads of this size have estimated development times (24–7 drilling, completion, and flowback operations) of approximately 20 months total with *associated truck traffic of 55–108 round trips per day in that timeframe, plus ongoing 23 truck trips daily during the wells' production lifetime.*"[28] Typical hydraulic fracturing processes create such substantial truck traffic because semitrucks had to haul in the massive amounts of water, sand, and chemicals to be pumped downwell at each site. This happens multiple times per well, given that the hydraulic fracturing takes place in discrete stages, rather than all at once.

In contrast, Marie's "stim center" was one central location where technicians could pump the water, sand, and chemicals into one steel pipeline that was connected to multiple wells. This meant that, rather than trucks having to make multiple trips to each dispersed well, they could visit one stim center and direct the flow of water, sand, and chemicals to whichever well needed to be fracked. This concentrated the main aboveground activity at one site that they could locate far away from neighborhoods, businesses, and heavily trafficked roads. Safety was an obvious concern, given the long length of pipeline—up to about two miles—traversed by the materials under high pressure. Anticipating that critique, Marie pointed out that the steel pipe was manufactured to the same specification of structural competency as the pipe used on the ocean floor in the Gulf of Mexico and said that they instituted and enforced a fifty-foot "red zone" to restrict personnel from entering the area when the pipe was pressurized and utilized pipe restraints made with bullet-proof Kevlar.

All of these design decisions could be justified in terms of efficiency and economics, but for Marie there was something more: stim centers were the morally correct thing to do because they lessened impacts for nearby residents while still providing an economic benefit for them in the form of taxes and royalties. She distinguished her company's approach by saying that the others who had attempted them did so for "purely efficiency reasons, to where they could bring a bunch of stuff under a central location and lower their costs and their disturbance footprint. You were saving costs on building locations, on roads, bringing in water, things like that." In contrast, she said, using the *we* of her company,

> we wanted to also benefit from those cost efficiencies, but a lot of it was because we were butting up next to people's houses and things like that. So a stim center is definitely a tool in a toolbox that we can use. . . . I remember a specific time that it helped us get the municipal permit, because we were able to tell these residents, "We're not gonna be right next to your house. We're actually gonna be two thousand feet over here, half a mile away, so you're not gonna get the lights and the dust and the noise as much."

Here Marie acknowledged a moral case for the stim centers, framed by the social license to operate. Doing the "right thing" by minimizing impact on residents while providing them economic benefit also helped the company secure government permits and save money—a win-win, in her eyes.

This approach also underlined the other examples of socially responsible engineering described by Marie. She worked on a project testing Colorado's first "electric frack fleet" of engines, which used natural gas and electricity instead of diesel. She described the engines as being "quieter and 95 percent better on emissions" compared with the diesel ones. "You could see the huge benefits on the emissions, and then noise and light as well," she extolled. Finally, Marie was proud of her company's innovative water pipeline project. They had laid 150 miles of pipeline that could transport water for hydraulic fracturing directly to well sites, strategically choosing to operate where they could link well sites to the pipeline. Doing so meant that, by the time of our conversation, she claimed that the company had transported over 62 million barrels of water and had eliminated over fifty thousand truck trips and over 10 million miles of truck traffic. She said it also allowed them to reduce the size of the well pads by up to 25 percent.

Excited that these efforts had provided tangible benefits to both the company and communities, Marie said, "So those were cool projects because of where we were able to incorporate feedback that we were hearing from people, you know, just our operations being in the communities and able to somewhat pivot or help lower those impacts." But, as suggested by her use of the terms *somewhat* and *lower*, she also wrestled with the limitations of how much she and her coworkers could do to appease nearby residents. Echoing chapter 4's focus on distributed agency, she recalled heated debates between different teams on how to spend money and determine timelines, describing how some teams were more motivated to cut corners on community engagement to speed up bringing a well online. "So it does come down to, sometimes, a moral decision versus an economic decision," she said. "You always have to keep both in mind. It's like the little triangle: the time, price, and quality. There's always one that suffers." By positioning time, price, and quality as trade-offs that had to be weighed against one another, she engaged in a logic of commensurability that others might seriously question or reject.[29] For example, the most fervent critics of fracking were not willing to sacrifice environmental well-being for increased efficiency.

In wrestling through those kinds of tough decisions, Marie said, she went back and forth between putting herself in the shoes of the residents and the shoes of her colleagues who were pushing for cheaper and faster community engagement activities. "It is difficult because you're working for [the company] and you're a [company] employee, but at the same time, I can relate to these people [community members], too, and, the impacts that they . . ." Her voice trailed off before continuing to express her empathy with the community relations group at her company, saying, "They're always in the middle, torn because it's like you're trying to do the best for these people that are being impacted. But you also have to think about the bottom line for [the company] and the cost of doing, you know, the extractive industry." It is telling that, when Marie seemed to reach the push point in which the costs of being responsible to impacted communities impinged too much on the company's financial bottom line, she symbolically handed over the dilemma to the stakeholder engagement team. While she said that she tried to put herself in the shoes of neighborhood residents, she acknowledged that because of the structural constraints of their work, most of the "listening" work fell to others.

STRUCTURAL BARRIERS TO LISTENING

Divisions of labor inside companies constrained most engineers' opportunities to listen to local residents, prompting them to devise other ways to pass along information among teams, as described by Marie. She pointed to the importance of drilling engineers—some of the first people on the ground—passing along information to the teams who would follow them. As completions engineers, she said, "We'd go to the drilling engineers and be like, 'Did you guys hear any feedback?'" She recalled that their responses would vary, from "Oh, they loved us, they brought us cookies" to "Watch out for this lady. She's very vocal and she needs some coddling and she needs some extra attention." In her company, they tried to formalize that feedback by recording it in the databases associated with each well. But she pointed out the limitations of the listening and change that could happen at the production stage of the process. "That's kind of in the reactionary space. We're already there, we're already impacting them," she explained. "So let's see what we can do right in that space. And so that communication was very good, and that's usually, like, putting up hay bales or putting up a sound wall or maybe trying to reroute our traffic." Here Marie acknowledged the limits of generating compatibility after the well facilities have already been designed and built.

Field-based experience provided some engineers with a lasting appreciation for understanding stakeholder perspectives, even as they moved into jobs that kept them at a desk. Kevin, the dedicated petroleum engineer profiled in chapter 4, spoke at length about the significance of his first years on the job as a production engineer for learning to listen to people outside the industry. As a typical entry-level position for petroleum engineers, the role involved enhancing the production of already existing wells. It required him to leave the office and visit his company's wells to assess and then implement mechanical, chemical, and other treatments to boost the well's production.[30] He always did so with one of the company's operators, a group of workers whom he described as "blue-collar men who live in the same towns where the wells are." Sitting together in a company pickup truck, he carefully observed how these men would interact with the people they encountered, including disgruntled landowners. "They just listened to them, kept their cool, and then promised to fix whatever complaint they had themselves or pass it along to someone who could," he remembered. When Kevin was promoted into reservoir engineering and helped plan when particular wells would be taken out of production, his previous experience taught him that those decisions had direct implications for his company's relationships with nearby residents. For example, he could opt to take an aging but not exhausted well out of production to improve the relationship with a local landowner. He recognized, however, that his work mostly kept him at his desk all day and that, like Marie, he was dependent on being "fed information" from the stakeholder engagement team.

Even engineers who recognized the importance of listening invoked "inside the fence" and "outside the fence" distinctions to set boundaries on which personnel "should" be responsible for interacting with communities.[31] Austin, who was the chief mining engineer for a large operation in Central America that was under intense international scrutiny for its troublesome human rights record (see chapter 2), also emphasized direct listening to community members to mitigate conflict. He underscored the importance of managing expectations and following through with promises but pointed out how shifts in personnel who cycled through projects during their careers made that difficult: "I learned that you have to build trust with the people. You have to follow through on it, you know, basically any promises that were made. Some of the problems we run into are that a lot of promises were made by people who no longer work here, and that generates a lot of mistrust when either you don't follow through or you can't follow through for financial or other reasons."

Austin's point shows how the time scales of mining projects and the distributed nature of the corporate "person" analyzed in chapters 4 and 5 make it difficult for people to hold companies as a whole accountable.[32] And even though he praised mining companies for becoming more "inclusive" in integrating stakeholder concerns into their operations, he also found it difficult to manage both his technical responsibilities and relationship building with stakeholders: "I'm really focused on the details of what goes on inside the mine gates. I obviously need to be kind of aware of all the issues that are going on outside. But there is always a wall between those two areas. Not because they don't want to communicate. It's just, there's so much to do inside with the technical part that I can't really get too involved with those other things." Even though Austin expressed a desire and a need to understand what was happening "outside" the mine gates, he found it difficult to do so while staying on top of the technical work he was formally tasked with assigning. He critiqued the institutional "wall"— work assignments, reporting structures, disciplinary teams—that seemed to artificially separate the mine from the world of social concerns "outside" of it, implicitly recognizing that the thoughts, feelings, and activities of people in the nearby villages would affect the mine's daily operation, and vice versa. But he also reinscribed the technical/social dualism by signaling the "inside" as being the "technical" domain that was his responsibility.

WHEN WIN-WINS ARE NOT POSSIBLE

The engineers I met held up shared infrastructure and design for community acceptance as aiming to maximize potential shared benefits between companies and their publics. In contrast, the limitations of the win-win proposition were explicitly recognized in compensation practices, since these acknowledged that not all harms could be designed out of the process and that affected people needed to be compensated for experiencing those harms. Anthropologists critique the underlying assumptions of compensation, arguing that it attempts to create commensurability between things that would otherwise circulate in different value regimes, for example, by proposing to replace a sacred and sentient glacier with trucked in water or replacing the loss of place-based livelihoods with cash.[33]

The engineers I met who were the most committed to greater public accountability viewed compensation as a last resort when other attempts to harmonize industrial activity with local concerns failed. Marie disparagingly called compensation "hush money." Aaron found placing a price tag on someone's complaint to be ethically troublesome, so he attempted to remedy the concern itself as much as possible. When speaking in my class, he illustrated his team's technique of "ask and listen" by describing their encounter with a vocal opponent of fracking near one of their operations. "When I actually sat down and talked with him, it turned out the thing he hated the most was that his car was continually getting dusty because of the dirt our guys were kicking up," Aaron said. "So I asked him if it would help if I arranged for him to have free premium car washes for the remainder of our time working on site." The neighbor agreed, and Aaron went to a locally owned carwash and purchased a punch pass for the man to use at his leisure, explaining, "There you go, he had his problem solved and we were able to support a local business at the same time." Aaron was careful to point out that even this "reactionary" sphere of listening could still improve future projects if it was fed back to the "beginning of the cycle when you plan the next project."

John saw similar limitations and opportunities of compensation from his position shaping the field of stakeholder engagement and performance standard compliance. After graduating in the mid-1970s with an undergraduate degree in mining engineering, he began his career in the burgeoning field of environmental remediation. He traveled the world, working as a contractor on large headline-grabbing projects before eventually accepting a full-time job with one of the oil and gas majors, where he specialized in strategic but controversial new international projects. He was in that position when the field of social and environmental sustainability reporting surged in the wake of the 1987 Brundtland Commission, which defined and set a globally influential agenda for *sustainable development*, and the 1990s transparency boom.[34] For the projects he worked on, he took on the role of navigating a host of new and evolving performance standards, most notably from the World Bank and International Finance Corporation. As of 2020, those standards covered eight key areas: risk management, labor, resource efficiency, community, land resettlement, biodiversity, indigenous people, and cultural heritage.[35] John had to make sure their projects met those standards; otherwise, the company would risk losing its funding.

Like Aaron, John discovered that he had to do considerable work inside of his company to convince his coworkers and managers of the importance of the performance standards. He said that it took major financial losses to eventually push the company's engineers—and the engineering-dominated management teams—to incorporate listening to local communities into the planning process. To illustrate, he described a multibillion dollar project to build new production fields, transportation facilities, and a plant in the South Pacific. In the following interview excerpts, he refers to his team by using *we* and refers to the engineers and other top managers as *they*, vividly underscoring the different and distributed agencies that make up corporate forms.

> The engineers in Houston looked at topo[graphic] maps and they had all the satellite imagery and the geotech, and they thought, "Oh, here's a great piece of ground. It's in the right place. We're going to put the gas plant right here and have design firms do it and whatnot." They drew the right-of-ways through the terrain for the pipelines based on purely technical criteria, all of the classical engineering conditions. What's the ground like? Is it a nonslip zone? Is it level? Where is it situated? So we got in the country, started doing

the surveys, and realized the pipelines were going through hunting grounds and sacred areas. We said, "Oh, we're going to have to move fifty families. It'd be easier to move the gas plant." They said, "Oh, no we've already invested the time and money in the design. So you'll just have to move the families. Oh, by the way we're going to start in a couple months."

Here John criticized the design fixation on the part of the engineers and invoked the "cost of conflict" argument used by many CSR specialists to shore up support for their work. The engineers' unwillingness to change the pipeline pathways ended up causing "more than eight months of delay and millions of dollars in compensation payments" for the families they had to resettle. This process generated considerable social unrest on the ground and eventually sparked what he called a "shift in design" inside the company, which involved substantial effort to overcome obduracy in how the engineers and managers planned infrastructure.[36]

> It was recognized that the transactional delays and the transactional costs were much, much greater than just a simple engineering decision, "Yeah, we can shift the pipeline one hundred meters this way and boom off we go." From that point going forward as they were routing the pipeline through the countryside, the engineering team took great pains to avoid residences and gardens and that sort of stuff wherever they could. Rather than just rely strictly on the right of way, we had teams out in advance talking to communities. "What are your sacred grounds? What's important? Which are the best marking areas? What is important to you, and whatnot?" So we were moving that right of way based on consultations with communities, as opposed to, "Okay, we're just gonna put it here because technically this is the right answer." Because, you know, they had to learn through the hard way that, in a traditional society, moving residences and moving people can be extremely difficult.

John's early prediction of the difficulties resettlement would pose for the project stemmed from the knowledge he gained from listening to local people and reading anthropological research on the area. He learned that they had a different relationship with the land than he and his North American coworkers: they recognized multiple and overlapping "use" rights embedded in a complex and malleable kinship system. Local people also viewed parts of the land as being home to their ancestors, even if their bodies were not physically buried there. These factors made compensation difficult and underscored the importance of what John called "active listening" and an "active feedback loop."

> What's the purpose of having stakeholder engagement if it's only a one way conversation, and you don't want to take feedback? Admittedly, some of that feedback might be painful, but, to me, that's the heart of it. You need to have that active feedback loop to go back to people and say, "Okay. We heard your complaint. We're going to do something about it. Here's what we're going to do. It's going to take this length of time." That communication process, that grievance management process, needs to be a robust, active, continually working cycle forwards and backwards. You've got to take it, and you've got to respond to it, and then you've got to live up to your commitments. If you tell people you're going to do something, then you've got to do it. Trucks driving too fast? Well, we'll go work with the project teams, and we'll get guys out there with flags. We'll do something to slow down the trucks. Kids can't cross safely to school? Okay. We can fix that, right? They may say, "We don't like the dust the vehicles are generating." Okay. That we can work around. "Here's what we're going to do. What do you think about this?" You need to take a look at what are the things that are due to us—and when I say *us*, the project—and how do we go about improving it?

In this narrative, John used the *us* to refer to a project in which the company's employees and contractors were united in making their operations accountable to the people they impacted. His years of experience on controversial projects convinced him that the public disturbances and protests were almost always "the result of a grievance that hasn't been answered." In so doing, he placed listening and response at the heart of companies' ultimate financial security—as long as that listening could generate feedback that they could pragmatically act on.

PRAGMATISM AND "ACTIONABLE FEEDBACK"

Challenging a persistent technical/social dualism, Aaron became convinced that the success of the stakeholder relations team rested on changing how their company's engineers and managers planned and executed their work. He and his team saw that leaving the community relations work until after wells were already drilled made it difficult and ineffective to address people's concerns, since there was very little they could do to change the practice of drilling or the well itself in response to the concerns they discovered.[37] They therefore professionalized a listening function inside of the company that aimed to integrate social concerns into the planning of specific wells and entire fields, as well as their everyday operation and eventual plugging and abandoning. Aaron valued listening as a tool to generate feedback: "There has to be listening, otherwise there is no feedback and no change."

The team developed multiple tools for this listening and feedback. In the areas closest to their potential operations, they went door to door to talk with as many people as possible. They invited a larger radius of people to neighborhood meetings that would allow more people to speak and ask questions than the typical kind of public hearing required by state and federal law. They made themselves available at community events. They created a response line that they personally staffed, categorizing calls according to type and tracking the resolution of complaints. They then developed creative techniques for translating the wealth of information they gathered from these sources into a form that was "actionable" by other teams inside of the company. These included maps of complaint locations, charts showing the time of day/night of calls to their grievance line, charts distinguishing types of calls into their hotline (e.g., complaints vs. requests for information), and GIS layers of schools, hospitals, and other key places that should be avoided when siting wells. Aaron explained:

> If we went and spoke narrative about someone who couldn't sleep at night, there's nothing that a drilling engineer can do with a narrative. . . . They don't become more efficient drilling a well from narrative. They don't control costs by narrative. Nothing happens in their world in just talking. They have to look at data, analyze it, and then make an action to it. It's the exact same thing. So we didn't have a narrative around the noise of rig activities at night. We showed a graph of complaints and time of day and it said, "The common element in this is you are delivering the steel pipe at three o'clock in the morning. Do we have to do that?" [They would reply] "Well, no. We could deliver

it at three o'clock in the afternoon." You could've spun that narrative and it wouldn't have mattered. It wouldn't have sunk in and attached to people. When you showed them the data, and then provided the context, you nailed it. Or if you showed the map exhibit and your rig A over here has these complaints, and rig B has no complaints, here's the proof that it's related to rig A and it's not some other company's rig. It's your rig and it's this specific item.

When Aaron said, "It's the same exact thing," he was referring to his team's ability to turn narrative into data that could be understood and engaged by people more skilled and comfortable with quantitative information. These tools helped engineers see the patterns and common themes in the narratives that the stakeholder team heard day in and day out. By creating graphs and charts, his team was able to carve a space for engineers to change their professional practice in ways that improved stakeholder relationships. This translation, however, was asymmetrical. To orchestrate an alignment of views, Aaron had to make the narrative information gathered by his team "speak the language" of their technical peers:

> That's the power of those exhibits and that's one of the ways that the technical person can participate in this is to help the socially oriented, the communications major, the sociologist, convert their world, translate their world into something that the oil and gas operative, technical person, blue collar team member can do something with. . . . The power of a map isn't that all the dots are accounted for. The power of a graph isn't that the trend is up. The power is that all of us who are looking at it have the same conclusion of what is going on. We all agree that those are the sum total of the dots.

One of the reasons this approach was compelling is that it illuminated specific causes of social problems that might otherwise appear irrational, unpredictable, or arbitrary, allowing both the stakeholder and engineering teams to formulate solutions that were more likely to result in good stakeholder relationships. This approach resonates with root cause analysis, a familiar exercise to engineers accustomed to identifying the causes of events like equipment failure. According to Aaron, "The power for the internal people is to dissect, 'Why did that generate that, and can we be aware of that as we go to plan the next piece, the next step? Can we be better when we plan the next step?'" The idea that there is a discoverable, underlying pattern of community conflicts bolstered their work because it suggested that such conflicts can be prevented in the future.

This stakeholder engagement group went on to become one of the most effective and emulated in Colorado for the positive working relationships they developed both outside and inside their company. But not even they could fully address each of the community concerns that came to their attention through listening, because not every complaint could be translated into actionable feedback. Given the wide range of opinions, concerns, hopes, and fears surrounding oil and gas production, they could not meet every person's expectation of them and their companies. A geological engineer named Ryan, for example, poured his soul into trying to address the complaints he received when he answered the company response line or met people in the town. Yet he also said that sometimes his hands were tied and the best thing he could offer upset people was acknowledgment of their criticism:

> When folks are in a situation like that, one of the things that they do is just let it out. Being able to just sit there and listen is huge. If I'm at a booth, I'm at a booth that says [my company name] and I'm wearing a shirt, and so it's clear that I represent the industry. I answer the phone, and I am a representative of oil and gas to them that they have the attention of. Some people—and I don't fault them at all for it, I certainly am one to do it myself at times—but they will take that opportunity and just talk and let everything out. It's my job to, as quickly and as efficiently as I can, help them in whatever way I can, and I personally find that tremendously rewarding.

Ryan found personal fulfillment in enacting the corporate form as a sympathetic listener, and he tried to influence the other agencies that made up that that same corporate person to address the criticisms and requests he received. But he also recognized that his own ability to truly address their concerns was limited by his structural position working for the company. He had to be pragmatic; he could listen at length but act only inasmuch as he did not hinder his company's ability to continue operating and drilling oil and gas wells. His experiences show how the constraints of the corporate context of their work generated "overflows" of perspectives and critiques of resource production that could not be contained within the category of actionable feedback.[38]

CONCLUSION

The engineers described in this chapter endeavored to make themselves, their companies, and their industries more accountable to multiple publics by creating "solutions" they considered to be mutually beneficial: those that would address residents' concerns while still generating financial benefits for companies and communities. I propose viewing this orientation to their work as pragmatic in nature, as it seeks to harmonize accountabilities to the public with their accountabilities to generate profit for the corporate forms employing them. In their accounts, the engineers positioned socially responsive technology—from stim centers to responsibly sited pipelines—as a source of industry progress in addressing the problem of community acceptance of resource production. Engineers such as Marie recognized their and their colleagues' agency in developing, testing, and implementing those technologies, even as they simultaneously pointed to the structural limitations of that agency: very few engineers had opportunities to interface directly with the people affected by their company's activities.[39] Even those that did, like Ryan, could not always act on the feedback gained through listening because they had to maintain rather than curtail industrial development.

But what kind of agency was this? First, it was *distributed*. The engineers had to work with others and work with others in mind. In addition to listening externally to nearby residents, they had to listen internally to their own coworkers and managers to learn how to advocate most effectively for the plans they were proposing. They tried to create alignment among the multiple agencies they encountered by creating actionable feedback that was more likely to be respected and taken up by their coworkers. This meant translating a wide array of public concerns, fears, and hopes into "data" that was more readily legible by other engineers, such as graphs and maps, and then situating this data within internally politically efficacious ideological frameworks, such as the business case for the social license to operate. While the engineers I met tried to influence the agencies of others to align with their own senses of what was right or best, they could not always force their coworkers or persuade their supervisors to follow their wishes. This limitation provides fodder for those who argue that the bureaucratic nature of engineers' work can disperse accountability: if everyone is accountable in some way, then no one is.[40]

Second, this agency was *pragmatic*. While the engineers were sympathetic listeners, their visions of reform were grounded in spaces of compatibility between the corporate forms employing them and the communities affected by them. They tended to define "wins" for both their employers and the public in financial terms. A senior petroleum engineer illustrated this point especially clearly. He prefaced his comments on social responsibility in the interview by asserting that mineral rights take legal precedence over surface rights, signaling that companies have more of a right to develop subsurface resources than surface owners have to prohibit it. Companies should not exercise its legal right, he said, to "force their way onto somebody's land" but "have a discussion with the people about what's going to go on, listen to their thoughts, their needs, their wants, their desires, and really come to a win-win." Imagining himself speaking to a resident, he said, "There will be a lease payment, and then royalties if there's a well on your land. And we're going to be paying taxes to your school district, so the schools should be better. We can improve the roads. You'll have some income."

This economic view of win-wins may appeal to those with the most to gain from it but does not acknowledge that others may differently define what a good life is, including by foregoing the economic benefits of industrial development to safeguard against potential social and environmental risks. When engineers have more power to define what a "win" is, it positions them as the developers of solutions *for* the people impacted by their work, rather than *with* them.[41] While a few of my interlocutors acknowledged the value of participatory environmental monitoring for increasing trust in industry, for example, they seemed to assume that "empowering" citizens to do science would ultimately vindicate the company against false accusations of harm—a far cry from activists who advocated for community-based research to make visible the harms of industrial activity concealed by industry-sponsored science.[42] One of the few engineers who seriously questioned the structures of expertise and authority that privileged engineers was Sofia, who found that she could not maintain that questioning and find peace with her job inside of industry (see chapter 4).

Questioning resource production itself played a minor role in this configuration of engineers' agencies. This is likely because this agency must be expressed through corporate forms that are financially invested in posing the question of natural resource production as one of *how* rather than *if* (see chapter 3), using the ethic of material provisioning as a justification for their activities (see chapter 2). The most radical questioning my interlocutors engaged in was whether, in certain times or places, natural resource production should not take place at all. As much as Aaron sought to make his company's oil and gas operations "compatible" with Denver suburbs, for example, he strongly opposed placing wells near schools and hospitals, and he advocated for the largest setbacks possible between potential wells and neighborhoods. Even the petroleum engineer who spent most of our interview talking about philanthropy (see note 14 of this chapter), thus striking me as firmly situated in the "old" CSR camp, surprised me by saying that her family's private company specifically avoided operating close to neighborhoods. "We've always tried to develop areas or drill on areas that are not developed, that are not subdivisions, as nobody really wants an oil well or a wellhead in their backyard," she said. "It was kind of our philosophy that we really didn't want to be there either because we didn't want it in our backyards. Why would they?" While the engineers' own aspirations of avoiding development where it is unwanted are laudable, it is also true that these industries as a whole have a highly uneven track record, along with national governments, of respecting people's rights to refuse natural resource production entirely.[43]

This pragmatic orientation of engineers to their work exists in tension with growing calls to radically change structures of natural resource production, consumption, and waste in the face of the accelerating climate crisis. Chapter 7 takes up this tension in detail.

"Engineers are the ones who implement a company's values," Bill explained to me over coffee, after I asked him why young engineers should care about corporate social responsibility (CSR). He had graduated with a degree in geophysical engineering from the Colorado School of Mines in the 1970s but then went to law school because he thought that the jobs available to him as a recent graduate had a too narrow technical scope. He went on to become an executive for two major mining companies, where he led changes in how they engaged both their critics and nearby communities. He collaborated with environmental groups to achieve a landmark land swap in the American West, designating some areas as wilderness to be protected from mining while allowing mining to proceed in others. He directed a greater share of the economic benefits from mining to the First Nations communities closest to their Canadian operations, including by establishing a local hiring policy and setting up a mentorship program for small businesses aspiring to become mine contractors. The activities he enjoyed recounting seemed very distant from those of the rank-and-file engineers I was eager to understand, which was what provoked my question to him. Bill continued:

> CEOs aren't the ones who are implementing a company's values—they just talk. Mine managers aren't the ones doing it—they just manage. It's the engineers and the truck drivers and the technicians who actually live out those values in the decisions they make: what kind of trucks they're using and how they are driving them, what schedules people are working, how the pit is designed,

which contractors they're hiring, how they treat the environment, all that. My job as CEO was to provide and express the mission, vision, and values of the company, but I'm not the one actually doing the company business.

Bill emphasized the agency of employees to enact corporate forms and, in so doing, positioned corporate accountability as emergent through the everyday practices of the people who bring these entities to life.

Bill's view of the corporate form resonates with the overall argument of this book: to understand the accountability of technoscientific corporations, we must understand the agencies of the people who constitute them. While multiple agents enact corporate forms, I have focused on engineers to show how engineering decision making is central to a corporation's overall public accountability. This book's ethnography points to the unique challenge facing engineers and the people who study them: engineers have unique expertise and capacities to act in contemporary controversies and debates, but they must act through others enmeshed in composite corporate forms.

To address this challenge, and the puzzles it creates, in this book I have proposed a new framework for understanding engineers' agencies in the context of corporate work. Engineers are not supercharged, individually autonomous agents, free to shape the world in their image, as both their most fervent fans and their harshest critics propose. But neither are they conformists who seamlessly embody insatiable corporate drives for profit. Engineers act as instances of corporate forms while also standing apart from them.[1] I have argued that the corporate context lends a distributed and pragmatic quality to their everyday practices of accountability: these are distributed because engineers are not sole authors of their actions and have to work through others, and they are pragmatic because they try to reconcile competing domains of accountability.

BETWEEN REFORM AND REVOLUTION

This book has traced the practices through which engineers encountered and attempted to reconcile contradictions between and within four domains of accountability: formal standards and corporate policies, their profession, the publics that cohered around their work, and their own ethical frameworks. They received no formal guidance on how to manage competing obligations, such as between serving as a faithful agent of their employer and protecting the safety, health, and welfare of the public. This lacuna left them on their own to reconcile those domains: they evaluated potential employers and clients for "fit" with their own values; tried to align the multiple agencies they encountered in the workplace under a goal of financially profitable social acceptance; devised what they considered to be win-win solutions that created financial windfalls for their companies and some of the people affected by their work; and enacted their companies as responsible, good neighbors as they spoke on behalf of those forms to others, from school children to family members.

When interpreted in the vein of Laura Nader's critique of harmony ideologies, these alignment-seeking activities could be interpreted as a "search for balance that domesticates conflict and often reproduces the status quo."[2] The engineers were searching for balance and desired to quell conflict, and their activities reproduced many parts of the status quo—chiefly, the continuation of mining and oil and gas production. But entrenching ourselves in the "hermeneutics of suspicion"[3] and solely emphasizing the conservative nature of these projects can make it difficult to grasp—and ultimately intervene in—how engineers themselves understand and practice their profession.

The engineers I met actively shaped corporate forms as they tried to craft themselves as moral actors. They created new institutional structures, developed new sociotechnical systems, and influenced their coworkers and supervisors to make decision making more responsive to public concerns. On a daily basis, they managed the trade-offs of particular planning, design, and operations decisions, or what Gary Downey calls technical mediation.[4] This required them, to greater and lesser extents, to redefine the problems posed to them. They engaged in perspective taking, trying to see those problems and the solutions from other points of view. Scott, for example, redefined the "problem" of drilling pad decommissioning to encompass the safety of the people he witnessed constructing informal settlements on them (chapter 5). After listening to fracking protestors, Aaron redefined the problem of the fracking controversy to be a lack of industry comprehension of and respect for public concerns, not the public's lack of information (chapter 6). Jen redefined the problem of resource production as how to reduce poverty rather than simply how to generate revenue (chapter 5). Attempting to "get out of [her] own skin," June redefined the problem of mine pit design to include avoidance of sacred cultural sites, not simply maximum economic efficiency (chapter 2). In each of these cases, the engineers raised the profile of social acceptance in the practice of engineering decision making.

These were attempts to create a composite corporate "we" they could live with and live *through*, since many of their professional agencies were expressed through distributed corporate forms. When my interlocutors found some alignment among these multiple accountabilities, they embraced and embodied corporate forms, speaking with the *we* of the corporate "person." When the limits of their own abilities to reconcile a multitude of public, corporate, personal, and professional accountabilities became evident, they detached from those corporate forms. Some engineers symbolically handed thorny "social acceptance" questions over to stakeholder relations teams, retreating to dominant professional ideals of engineering as technical rather than sociotechnical in nature. Others emotionally checked out of their work without leaving their jobs, dutifully showing up to work but describing those collective corporate forms as "they." This subtle but powerful shift in language distanced themselves and their own accountabilities from the larger corporate forms, in effect disavowing themselves of responsibility for the actions of the others with whom they did not agree. Others sought more professional autonomy as consultants, and a few left industry entirely.

While writing, I found myself wavering between describing these efforts to practice more socially and environmentally responsive engineering as projects of *reform* or as projects of *transformation*. Calling them projects of reform seemed to emphasize what stayed the same around them, namely, the broader institutional constraints of engineering work in for-profit companies. In contrast, calling them projects of transformation seemed to index my interlocutors' sense of the significance or magnitude of the changes they sought. For example, one could describe Aaron's efforts to change engineering practice to manage the fracking boom (see chapters 1, 6) as a project to either reform or transform. He saw that integrating the social license to operate (SLO) into planning, operations, and engagement would generate greater compatibility with the growing suburbs, allowing his company to maintain or expand its ability to drill new wells in the midst of the fracking controversy—reform. But creating those changes required him to challenge dominant institutional practices and assumptions about everything from the deficit model of public engagement to the exclusion of "social" disciplines from planning decisions—transform.

Using the analytic lens of reform draws attention to how these projects ultimately reinforced the power of corporations—a persistent concern for critical social scientists. By shaping public debates about mining to be about *how* rather than *whether* to mine responsibly, Stan Dempsey and Art Biddle were able to secure permits for controversial projects (chapter 3). The engineers who consciously enacted the corporate person as friendly and responsible became a kind of public relations army in the battle of public opinion (chapter 4). The consultants who tried to make their corporate clients more responsible shored up those clients' reputations and claims to legitimacy (chapter 5). The engineers who created "compatibility" between industrial activity and suburbs maintained their employers' ability to work in socially and environmentally complex contexts (chapter 6). In all these practices of accountability, engineers who invoked the SLO added community acceptance as a key factor in managing risks to ensure profitability. These were pragmatic politics that sought to make engineering more accountable to the public while safeguarding their employers' place within existing systems of natural resource production.

DIVISIONS OF LABOR AND AUTHORITY

The engineers' efforts to reconcile competing accountabilities—and perhaps to change corporate practices—were complicated by the distributed nature of the agencies they could embody at work. They were, in very tangible ways, agents of others and acted through others. But although the engineers' agency was mutually imbricated with others, not all agencies enjoyed the same authority. Workplace hierarchies meant that those who rose through the ranks to attain positions as managers and executives, such as Dempsey and Biddle, had more social and financial capital to set the "tone at the top" and align the disparate enactors of the corporate form around a notion of financially prudent social responsibility. While these men led institutional transformations aimed at greater accountability to a broader section of the public, many of my interlocutors perceived the people at the top of their chain of command—the executives and boards of directors—as too narrowly accountable to shareholders and the financial bottom line. Engineers like Aaron and John (who specialized in performance standard compliance for the oil and gas industry; see chapters 2 and 6), for example, had to work laterally and from the bottom up to gain traction for projects aiming to improve social well-being, and they invoked the cost of community conflict to win over hard-nosed managers. Consultants such as Scott and Peter (who set up cooperative research agreements between companies and government agencies; see chapter 5) had to engage in careful diplomacy with their corporate clients, advocating for what they believed to be the most accountable course of action without endangering future contracts.

The division of labor inside of complex corporate forms also shaped engineers' opportunities to listen to members of the public and act on what they learned in those interactions. My interlocutors who were the most committed to improving the public accountability of their industries all described formative early experiences listening to and learning from people affected by their industry.[5] Those who ran their own consultancy firms maintained more direct engagement with people on the ground. Those who stayed in corporate jobs found themselves increasingly tied to an office desk, without opportunities to directly engage with stakeholders themselves outside of occasional volunteer activities. This is partially because even the most forward-looking companies institutionally separated "social" and "technical" functions for all but senior personnel and major decisions.

This technical/social dualism was a powerful ideological force even for those who fought against it, such as Jen, who wanted to ask questions about how changes in underground mining methods would affect a mine's social contract with workers and communities. It was precisely an unsatisfied desire to ask questions about the inherent social dimensions of engineering practice that prompted people such as Bill, Art Biddle (chapter 3), Addie (chapter 4), and Lila and Jen (chapter 5) to obtain graduate degrees in law and development. Their experiences add to a growing concern that an endemic "culture of disengagement" diminishes undergraduate engineering students' commitments to public welfare and pushes out the students who express the most interest in social responsibility.[6] These trends are worrisome and deepen a mismatch between undergraduate and professional experiences, given that working on sociotechnical challenges is a source of career satisfaction for large numbers of practicing engineers.[7]

WHAT PROMPTS ENGINEERS TO EXPAND THEIR ACCOUNTABILITY TO OTHERS?

Given the constraints posed by engineers' undergraduate training, their professional socialization, and their work in corporate contexts, we must ask the question, *What prompts engineers to step outside of their own positionality, subjectivity, and expertise to expand or enhance their accountability to others?* This move was critical for the practices of accountability analyzed in this book, from Dempsey and his engineering colleagues learning from ecologists in the 1960s to Aaron and his team learning from residents about what it was like to live in close proximity to oil and gas operations in the 2010s. This perspective taking is not trivial, given that many engineering codes of ethics explicitly argue that engineers should "perform services only in the areas of their competence."[8] The code could be interpreted to warn engineers against taking on roles for which they received little or no formal training, such as community engagement. Yet accountability requires a person or group of people to whom to be accountable, and engineering decisions cannot be made more accountable to those people without learning who they are and what they hope, fear, and desire. The perspective taking I saw engineers doing in my research was indispensable for that learning about those others. Crucially, it invited self-reflection on engineering knowledge, repositioning it as professionally and culturally specific ways of knowing and valuing the world rather than as the universal way the world ought to be understood.

Weaving through the ethnography presented in this book were two key sources of motivation for perspective taking. The first was what engineers described as intrinsic desires to understand and help people who were less fortunate than them.[9] For example, Scott, the consultant who used his geological engineering expertise to make roads safer and more socially just, pondered the question of accountability by reflecting, "The engineering code does not drive me to want to consider social issues, nor does any other code, I guess. So where does that come from? In the core of somebody, where? What type of person is this and why?" He then pointed to his own experiences growing up, "going to church every Sunday" and "every summer going on mission trips." Such service, he said, was not a "burden" or something he thought he "should" do but "a part of life . . . an inseparable part of who I am." This emphasis on service echoed through many engineers' narrations of their life and career histories, who said they always wanted to use engineering to benefit the people who were most in need.

Crisis management was a second and distinct key source of motivation for engineers to step outside their own positionality, subjectivity, and expertise. This framing was reinforced when the SLO was the dominant framework used to understand and act on public accountability. Engineers described consciously seeking out conversations with the people who were most critical of their industry as a way to understand and try to mitigate public outcry. Exemplars of this practice were Aaron, who attended public meetings, went door-to-door in the neighborhoods located close to where his company was seeking to operate, and devised strategies to communicate stakeholder concerns to operations teams; and John, who worked to assure performance standard compliance for controversial oil and gas projects around the world. Other engineers listened to stakeholders in more superficial ways and struggled to use what they learned from that listening to inform engineering practice. While the crisis management frame invited engineers to more accurately understand people's criticism of their industry, it can make the case for greater accountability by portraying people as potential risks to managed. Scholars caution that this framing can exist at cross-purposes with other goals of community engagement that require trust-building, such as sustainable development, or with broader challenges, such as grappling with climate change.[10]

WHAT PROMPTS ENGINEERS TO UNDERSTAND THE ENVIRONMENT AS MORE THAN RESOURCES?

The unique challenges and opportunities posed by work in the resource industries in the age of the Anthropocene raise this more specific version of the question posed above about engineers stepping outside of their own positionality, subjectivity, and expertise to expand or enhance their accountability to others (figure 7.1). Anthropologists have long argued that notions of a bounded "nature" separate from "society" are a modernist conceit, suggesting

Figure 7.1
Utah's Bingham Canyon copper mine, which is visible from space. Source: Spencer Musick via Wikimedia Commons.

instead that nature "partakes, but without being entirely, of the human."[11] Social scientists show that the things we call "resources" do not already exist in the world but must be brought into being through "complex arrangements of physical stuff, extractive infrastructures, calculative devices, discourses of the market and development, the nation and the corporation, everyday practices, and so on, that allow those substances to exist as resources."[12] Resource ontologies play a key role in these processes, as they ascribe future financial value to earthly materials and engender anticipatory affective states, such as hope and speculation.[13]

I was struck by how often my interlocutors stepped out of resource ontologies that valued nature as a source of economic value for humans. Most often, this happened through their own experiences enjoying the outdoors.[14] It was the love of wilderness, for example, that led to Scott and Gary (who founded his own non-profit organization; see chapter 5) to choose geology and geological engineering as undergraduates and fueled their own angst that their consulting work would facilitate the destruction of the very places they loved so much. Elijah, the geological engineer who left the oil and gas industry to volunteer and then return to graduate school (see chapter 4), directly tied his desire to think about "big R" responsibility questions, such as how the design of cities locked people into dependence on petroleum products, to his ties to the rock climbing community. He said that, caught up in his everyday work at the oil and gas company in Houston, it was easy to go along with the ethic of material provisioning, to find himself thinking, "Yeah, that makes sense, why not? We need all this stuff in our lives." But he said his friendships in the climbing community back home reminded him, even at a distance, that mountains had spiritual, social, and ecological value, rather than simply being resources for human consumption (figure 7.2).

In my research it was difficult, in fact, to find an engineer who did not profess some sort of love for the outdoors. This raises the question of what other work these appeals might have been performing in the interview context itself, given that many of my interlocutors perceived our exchanges as an opportunity to enact their own accountability. Recall that acknowledging that earthly substances and relationships are more than resources is also a pervasive technique of corporate representation, all the way back

Figure 7.2
Elijah, a geological engineer, enjoying the "other-than-resourceness" of the natural world, rock climbing in Colombia. Used by permission.

to when Art Biddle introduced Kay Ferrin as a "sensitive engineer" and "professional ski instructor" to a crowd that saw the potential Mt. Emmons mine as a threat to their town's efforts to brand itself as an outdoor tourist mecca, complete with a brand new ski resort (see chapter 3). The petroleum engineer River (chapter 4) experienced this envelopment into larger corporate performances of environmental virtue firsthand while working in Alaska. An avid backcountry skier, he saw the need to rebuild a warming hut in the mountains and received partial sponsorship from his company's philanthropy department for the project. At the time, the company's public relations push was to portray the company's employees as what he called "average Alaskans." River recalled:

So myself and four or five other employees donated over two hundred personal hours toward rebuilding a hut in the mountains. We did that out of our passion for seeing the old hut getting run down and [worrying] it was going to get torn down. So we wanted to help out. We did that and organized a grant in addition to our time and had our time matched in dollars from [the company]. And [the company] thought this was great and featured us in a big full-page holiday ad. There were three of us that ended up in this ad in the Sunday paper that highlighted the work we did on this mountain hut. It said what our job titles were and that we had a passion for the outdoors and we spent time working on this hut and [the company] was backing us and backing the 501c3 [nonprofit].

The ad turned on crafting an environmental reputation for the company as a whole by playing up the fact that its engineers—identified by their job titles—cared for the environment in ways that exceeded treating it as a resource. I witnessed a similar performance of other-than-resource environmental virtue in real time during a tour of a controversial gold mine in the Pacific Northwest in 2010. I was touring the mine with the head environmental engineer and the leader of the grassroots environmental organization that had nearly stopped the mine from opening. We were walking around the hilly wooded area where the company was proposing new drilling that the environmental group opposed.[15] The environmental engineer was affable and rosy-cheeked, dressed in a flannel shirt, jeans, and boots that suggested an easy transition from work into hiking. We heard a bird call, and the leader of the environmental group quickly identified it. The environmental engineer smiled but quickly identified the next call we heard, sparking a friendly-yet-terse exchange between the two men as to how many bird species they had identified in their many years of enjoying the outdoors. As each tried to outdo the other in his knowledge of the local environment, I was reminded of the innumerable CSR reports I had scoured that professed the company's commitment to environmental stewardship by showing its employees, clad in hardhats, out in nature.[16] It also reminded me of my visits to mining and oil and gas company offices and worksites, where I found walls filled with professional photographs of wildlife coexisting with the company's infrastructure, such as regal elk grazing next to conveyor belts back home in Wyoming.

These varied performances of environmental virtue, I propose, seek to send a message that engineers, in their roles as enactors of companies, are making uniquely judicious decisions about which parts of the natural world should become resources. These performances situate the engineers as "planetary managers"[17] inside a frame of their respect for the environment. Many of the engineers I met positioned themselves as an avid outdoorspeople who were distinct from their environmentalist friends in their knowledge that expensive outdoor performance gear as well as gas to power SUVs up to ski resorts depended on oil production. These acknowledgments of the environment as more than resources, therefore, ultimately shore up the authority of the larger corporate form by attributing the environmental virtues of employees to companies as a whole, again playing the scales between human beings and corporate persons (see chapter 4).

HOW COULD DECISIONS ABOUT EARTH RESOURCES BE MORE ATTENTIVE TO DIVERSE PERSPECTIVES?

Social scientists would point out the limited role for the public to shape the idealized judicious engineering decision making implied by these practices of environmental accountability. The "problem" of the kind of intense industrial development epitomized by mining and oil and gas development is frequently posed as one of "not enough" public participation. If only the public could have their voices heard by decision makers inside of corporations and governments, this line of reasoning goes, then they could influence or halt that development.

This approach, however, belies the fact that decades of "more" and "better" participation undertaken under the banner of CSR have failed to resolve conflicts between corporate forms and their publics. A key characteristic of the rise of CSR is the surge of tools used to channel increased demands for public participation: public hearings hosted by government bodies; community meetings hosted by private industry; social impact assessment conducted by consultants; grievance lines, text message complaint logging, and social media platforms administered by companies; sustainability reporting conducted by consultants on the behalf of companies;

and so on. Yet Andrew Barry's observations of the unprecedented "transparency" initiatives surrounding the early 2000s construction of a major oil pipeline hold true for these growing engagement strategies as a whole: rather than generating consensus, the explosion of information leads to new forms of dispute, concerns, sites, problems, and subject positions.[18]

Moreover, existing models of participation privilege corporate interests over those opposing them.[19] Chapter 3 shows that this outcome was explicitly desired by the first engineers-turned-lawyers who laid down the foundations for environmental impact assessment. Even less formal community meetings, such as those described in chapter 4, were strategically intended by their architects to forestall what the representatives of one oil and gas company called "soapboxing," in which commenters used the public comment period to give impassioned speeches opposing the project under question. A company representative once told me, as we watched residents mill around the tables at one of those meetings, that the one-on-one conversations "took away the platform" from staunch anti-fracking activists. And chapter 6 shows how the category of "actionable feedback" privileges matters of public concern that can coexist with industrial activity.

To confront the interlinked material, environmental, social, and technological challenges embedded in debates about resources, we need not just more public participation but the "capacity for disagreement" that does not simply shore up the power of the already privileged. Christopher Kelty argues that public participation tools are animated by particular "grammars" or "forms of life" that guide participants in understanding and judging the world. In its currently dominant form, participation "washes out the ability of different collectives to judge the world differently, and to participate in the clash of these judgments."[20] The ethic of material provisioning, for example, emerges from a worldview that defines the "problem" of natural resources to be one of increased production to meet increasing consumer demand. This view marginalizes or scorns other understandings and judgments of the world, such as those that center resource conservation rather than increased production.

The limitations of officially sanctioned participation fuel the fire of protests, which can be viewed as overflows outside of the officially sanctioned tools for public participation. Protest invites and provides positioning for citizens to critique the process of engagement itself rather than just become enveloped by it. This can make visible how public hearings, community meetings, and other CSR tools are weighted in favor of project proponents. Clashes can be a "motor of exploration, critique, and renewal,"[21] such as when the #NoDAPL movement defined the "problem" of the Dakota Access Pipeline to be one of protecting water and sacred ground, Native sovereignty, and climate change mitigation.

But clashes in how we understand and define the world can also lead to entrenchment that impedes translation. Debates about our resource futures too often slide into well-rehearsed jabs at the opposition that call out individuals for systemic problems. In the context of these conflicts, a criticism of industry can be intended or interpreted as an indictment of the morality of the people who work in it, prompting people to defend their own moral standing by vigorously defending an industry as a whole rather than reflecting critically on it.[22] This form of conflict heightens defensiveness and rigidity in thinking on the part of both proponents and opponents.

It is difficult to disagree when using different grammars; we may be able to understand the words spoken by others but be unable to make sense of them. A crucial but challenging aspiration, therefore, might be nurturing spaces and processes that help make different forms of life legible to one another, so that we can collectively reflect on the ones we use, why, and to what effects. For engineering students and professionals, this should include not just acknowledgment that different understandings of "public good" or "progress" exist but that these are animated by different understandings of what the world is: earthen materials are not just potential resources for human profit.

HOW CAN WE EXPAND THE FRAMEWORKS
OF ACCOUNTABILITY THAT GUIDE ENGINEERS?

This book has identified two dominant, interrelated moral architectures: the social license to operate (SLO) and the ethic of material provisioning (see chapter 2). I have suggested that their ubiquity is due to their seductive promise to reconcile otherwise competing accountabilities, such as between public welfare and corporate profit. Scholars seeking more creative imaginings of resource production, consumption, waste, and conservation can acknowledge and engage internal critiques of the dominant images of accountability available to engineers and put them into conversation with social science thinking and research.

Beyond the Social License to Operate

The SLO framework for thinking about accountability ostensibly raises the profile of community acceptance inside of corporations but nonetheless reinforces what John R. Owen and Deanna Kemp critique as an "industrial ethic" that hinges on narrow perceptions of a company's self-interest, chiefly those of production and cost.[23] The SLO marginalizes people and concerns that cannot be portrayed as constituting a significant business risk, and it channels engineers' practices of accountability into creating "win-win solutions" that financially benefit companies and some portions of the public. This book points to the limits of accountability that is justified by profitability: it leaves out accountabilities that may be ethically justified but not ultimately good for the financial bottom line. This should raise caution for the rapid uptake of the term *social license to operate* in other industrial sectors and public spaces, such as biotechnology.[24]

Engineers, too, critique the SLO. Emma, the petroleum engineer whose confrontation with landowners is described in chapter 4, called the SLO "gimmicky" while admitting that it was "smart business." Julie, an engineer who worked on major environmental remediation projects in both mining and oil, worried that distinguishing a SLO could feel like an "add-on or a differentiation, whereas it should be part of your core philosophy of operation; otherwise, it gets a little too much window dressing versus actually becoming a sustainable embedded concept in how you actually operate your facility." Even Aaron, who used the SLO concept to raise the profile of community acceptance in engineering decision making at the oil and gas company where he worked, said the SLO was the "bare minimum" and that companies had to think more expansively about their corporate responsibilities. And Art Biddle, an early architect of the business case for community acceptance, firmly believed that companies had to do more than argue they were producing jobs and tax revenue. Speaking to my CSR class, he said, "The jobs argument is the weakest one you can make. Money is a transactional relationship that evaporates once it's gone. The meaningful, enduring relationships are those in which you understand and respect people's hopes and dreams and visions for their lives, and then you see if you can help support those."

Engineers who took a more active role in stakeholder engagement activities in their work were quick to criticize the limitations of the SLO concept in light of the other, more formalized frameworks they preferred. Originally trained as a civil and geotechnical engineer, James relished the opportunities his career afforded him to "travel the world and meet people from different countries and different cultures." One of his first jobs impressed on him the importance of developing empathy with people who came from very different backgrounds. After graduating in the 1970s, he accepted a job contributing to the design of a nuclear power plant. As a part of determining ground motions, he traveled to remote parts of rural Europe to learn from locals about past earthquakes:

> Part of what we had to do was kind of meet and talk with some of the local villagers and kind of interview them from a historical basis of what earthquakes they recall and what had happened. And in doing that, we sometimes would go in their houses. This wasn't in the auditorium where we were interviewing them. We were interviewing them on the street or in their houses, almost in the neighborhood. It was a way of putting yourself in their shoes, seeing how they lived and talking to them. . . . That was the way that really helped developed some empathy for me, just being able to talk to these people locally in their language and kind of in their environment and therefore get the information or data we were looking for versus taking it off of a piece of paper or computer output.

James went on to a successful consulting career that grew alongside the 1990s explosion of reporting standards and social and environmental impact assessment, working on multiple high-profile international oil and gas projects. James stood out among my interviewees in his respect for local knowledge and desire to incorporate it into his engineering work. He used his experiences traveling to reflect critically on his own assumptions: "I think that [travel] helped me understand not everyone lives like we do in the US and wealth is quite different around the world and there are a lot of poor and indigenous people that in some cases are very happy indigenous people. So maybe that helped me in my engineering career to be much more perceptive or sensitive about different people and different cultures."

Through his work, James became well versed in the fields of impact assessment and sustainability reporting, which offered more formal alternatives to the dominant SLO discourse that emphasized a company's voluntary actions. For example, he recognized the complexity of attempting to establish free, prior and informed consent as recognized by the UN Declaration on the Rights of Indigenous Peoples. He rhetorically asked, "What is consent?" and then spoke to the internal power dynamics that created divisions and inequalities within "communities" that are made to appear homogeneous in the pages of glossy corporate reports. When James imagined talking with young engineers about the social dimensions of their careers, he invoked an image of learning as an intrinsically valuable "lifelong process." He said to me: "Be careful thinking you've learned it all. I'm about to turn sixty-five, and I get up every day and I feel challenged and I am going to learn something new today, whether it's from people like you or others. Education kind of helps launch you, but the key thing is to be open and committed to continually learning and improving." The SLO was absent from how he imagined talking with young engineers, in stark contrast to the narratives of the other engineers who emphasized the SLO to persuade novice engineers to care about social concerns (see chapter 2). Instead, James portrayed the point of his career to be one of continual learning about the world, including respecting different definitions of a "good life."

We must go above and beyond the SLO framing of accountability to one that is minimally based in free, prior, and informed consent and includes agreements that are continually monitored and enforced. While not perfect, this would provide the publics that are most affected by industrial projects more say in defining what "health, safety, and welfare" are to them, deliberating whether projects present acceptable or unacceptable risks to those values, and assessing companies' performance against their promises. Provisions would need to ensure that, within those publics, the most vulnerable people who stand to lose or risk the most are not overruled by a simple majority and that some form of government enforcement exists to hold corporations accountable when agreements are breeched or damages occur.

Yet even informed consent, as Dempsey and Biddle recognized fifty years ago, is a weak form of public participation, as it asks the public to approve or disapprove already designed engineering projects. A far more meaningful notion of accountability would invite external critique and modification in the planning process itself, rather than giving publics the option to oppose or affirm an already mature proposal. For that kind of participation to be meaningful rather than simply co-opting, companies would have to be ready to walk away from projects if their publics ask them to do so. And before such planning even begins, more robust accountability would acknowledge different judgments of how the world can be "good." Rather than asking how mining can be done responsibly, it would ask, *Can mining support citizens' desires for their future, and with what potential risks and safeguards?* Rather than assuming shared purpose, it would investigate whether and how shared purpose is possible.

Beyond the Ethic of Material Provisioning

The ethic of material provisioning is more historically durable than the SLO, animating generations of people who work in or grew up in towns that took pride in their mining or oil and gas histories.[25] It places profit generation within a much broader vision of material provisioning. The engineers I met positioned themselves as moral actors by emphasizing their unique ability to provide the material foundations of modern life: hospitals, transportation, communications technologies to connect and entertain people, and even renewable energy. They were quick to point out that even the enjoyment of nature—a space that is popularly associated in the United States with the absence of industry—depended on products made from oil, from contact lenses to outdoor performance gear.[26]

While this powerful image of material provisioning captures some truths about systems of resource production and consumption, the time has come to create other images of accountability and other ways engineers can contribute to a greater "public good." The ethic of material provisioning marginalizes engineers' other agentive capacities, such as critically questioning resource consumption rather than simply delivering more resource production. Moreover, it has failed to foster support for these controversial industries and the people who work in them, despite being the main message of industry public relations campaigns for at least a century. Even the growing explicit recognition among social scientists and humanists of our collective everyday reliance on the materials produced through mining and oil and gas development has not resulted in those scholars turning to praise the people who work in industry, as many industry boosters hope. Instead, academics use this increased recognition to radically reimagine ways of living that are less dependent on these resources, especially oil.[27]

There are already seeds of discontent. Chapters 5 and 7 shared the experiences of engineers who either critiqued the ethic of material provisioning or rejected it entirely. Most of these engineers were consultants or had left industry entirely. Peter, the environmental engineer who spent most of his career creating ties between mining companies and government agencies, worried about the overall sustainability of consumer society. He critiqued the mantra that "if you can't grow it, it has to be mined" as a "defense" that was true as long as consumer society remained as the status quo. He then invoked his own identity as an environmentalist, stating, "We're defensive against mining because people need it and people like their stuff. Those of us who are tree huggers say, 'Yeah, but we have too much stuff and we're not using our stuff wisely, and we're throwing all our stuff away.'" Here he redefined the problem of the material basis of consumption to be one of conservation rather than simply meeting increasing demand—an analytic move that resonates with scholars who criticize the senses of inevitability promulgated by industry appeals to increased production. Such thoughtful internal reflections are difficult to access when industry insiders feel personally blamed for systemic problems that involve but exceed their own agencies.

Upsetting the sense of inevitability proffered by the ethic of material provisioning would require serious consideration of the places where resource production ought to occur—and ought not occur. Some companies and personnel acknowledged that they did not want to operate where they were not welcomed, but this too often came across as a rhetorical strategy to enhance the company's caring image rather than an actual commitment. Geological engineer Elijah, for instance, said that when he was working in the oil and gas industry, "If you reached into the realm that directly went against the overall philosophy of what it means to extract oil and gas and why, that was entering a taboo realm." For petroleum engineer Emma, seeing a rig come in to drill wells close to her own home accelerated her own reflection on where such industrial activities should and should not occur:

> From our house you could see four well sites. . . . There was actual well site access from our neighborhood road and it didn't have a pumping unit on it or an engine, it was just a naturally flowing well and I never thought anything about it. But when the rig came in, it made me really uncomfortable. Why is that making me uncomfortable? It shouldn't make me any more uncomfortable because it's in my yard; I still believe in what we are doing and the integrity of it, but I didn't like it. It made me feel really weird. I didn't like it because it was in my backyard.

With enhanced empathy for the people who live near the well pads she designed and helped maintain, Emma used the experience to engage in perspective taking. "So I get how they are feeling," she said. "I wouldn't want a gas-fired engine running 24/7, 300 feet from my deck." Despite this moment of empathy, she also firmly advocated against companies going above and beyond the legal requirements to accommodate residents' concerns and desires (see chapter 4).

What would be conducive to more just and sustainable resource futures is a new ethic of relation, not an increasingly loud ethic of material provisioning. Though engineers celebrate that provisioning as a service, others can experience it as paternalism.[28] Engineers need to be accountable not to an impossibly nebulous notion of the public good but to particular publics. Debates and decisions about resources should hold space for diverse viewpoints, but we need new ways of managing what Kelty calls the "perplexity" that emerges when one form of life is illegible to others, or when our "grammar" does not help us interpret another's language: "Perplexity is not disagreement, it is rather the *inability to disagree*. . . . To share a form of life is to engage in resolving perplexity through shared judgments of the world. Without such shared judgments, the language of others makes no sense—even if one can understand it."[29] He goes on to suggest that such perplexity should be an invitation to recognize our own grammars rather than simply denigrating others:

> The perplexity we experience when, for instance, people refuse to believe in the existence of climate change should perhaps be similarly diagnostic for us, and not simply an occasion to pathologize or accuse (resentment, on the other hand, may be a common response to such denial, and a sign of this perplexity).[30]

Enhancing the accountability of large, technoscientific organizations, therefore, partially rests on opening up learning opportunities and institutional positions in which engineers can view their way of understanding the world as one among many, in which they can learn about how diverse publics define good lives, in which they and their critics can experience perplexity and, rather than point fingers, ask questions to start translating among those different judgments of the world. Engineers and their interlocutors all deserve more capacious questioning about engineering, resources, and society. These activities are present in engineering but are obscured by dominant images of the profession as technical problem solvers in the service of industry, animated by a vague notion of the public good.

WHAT ROLES CAN ACADEMICS PLAY IN REDEFINING DOMINANT IMAGES OF ENGINEERING?

In addition to encouraging us to ask other questions about the engineering profession, this book invites scholars and engineering educators to reconsider our own practices, such as with whom and how we conduct research, as well as how the outcomes of that research travel. As academics pursue scholarship that addresses pressing public problems, too often a bifurcation emerges in which we work in solidarity with people who already share our political sensibilities or we engage in a militant anthropology of elites predicated on morally insulting those who work in the industries we criticize. The growing urgency of climate change has exacerbated these tendencies, leading to a flourishing of scholarship that celebrates critique of the capitalist underpinnings of petromodernity.

What about those of us who listen professionally to people judged as ethically suspect by our colleagues, to people who seek a "kinder, gentler capitalism" rather than the "potential of capitalism's undoing"?[31] The book opened with Aaron feeling attacked by both anti-fracking activists and his own colleagues who suggested that "by *listening* to 'those people' we were sympathizing with them." I suspect Aaron's story resonated strongly with me because I felt similar dislocations when returning from the field of engineering practice and education to share my experiences with fellow anthropologists. How can we enter into our interlocutors' lives to document and theorize the ethical frameworks animating them while maintaining a critical eye to the inherent political dimensions of these frameworks that enable asymmetrical distributions of benefits and harms?[32] If we seek for our research to contribute in some way to addressing the challenges it identifies, how can we do better than bifurcating our writing into reformist reports for practitioners and more expansive reimaginings for fellow sympathetic academics?[33]

One concrete way that scholars and educators can critically participate in enhancing the accountability of engineers and the corporate forms they enact is to support engineers becoming new kinds of knowledge workers. To address the challenges of designing a more "sustainable knowledge infrastructure for our species," Geoffrey Bowker proposes two new kinds of knowledge workers: "brokers" who can "move across knowledge communities" and "transducers" who can "transform data, knowledge, and practice in one arena and prepare it for effective use in another."[34] The engineers I met who were most committed to public accountability routinely engaged in both of those practices in their work. They also regularly asked themselves questions that resonate with calls for more socially just engineering practices: "Why am I an engineer? For whose benefit do I work? What is the full measure of my moral and social responsibility?"[35] We should be accountable to cultivating and celebrating these qualities in engineers, especially during their professional socialization.

As detailed in the epilogue, I have engaged in critical participation as a platform from which to "reimagine the nature of knowledge for the way the world is now."[36] Engineering education and practice presents particularly fertile ground for this work. Carl Mitcham identifies what he calls a philosophical inadequacy of engineering. Though definitions of engineering are multiple, dominant ones involve technical knowledge grounded in a service ideal, understood as promoting "human use and convenience" or public safety, health, and welfare. "But there is nothing in engineering education or knowledge that contributes to any distinct competence in making judgments about what constitutes 'human use and convenience,'" Mitcham writes, let alone public safety, health, and welfare. Provocatively, he continues, "Engineering as a profession is analogous to what medicine might be if physicians had no expert knowledge of health or to law if attorneys knew nothing about justice."[37] If the long established engineering ideal of service is to remain, social scientists and humanists must work with engineers not to define, once and for all, what the "public good" *is* but, first, to think through the question of what could engineering be for and, second, to understand why people answer that question differently.[38]

Grappling with the growing urgencies posed by climate change and social, racial, and environmental injustice will require addressing the well-worn grooves of engineering pragmatism, as these impulses seem to exist at cross-purposes. Dominic Boyer, for example, calls for "revolutionary infrastructure," arguing that "an incremental, partial, slow transition away from fossil-fuel sources and infrastructures is simply not a luxury we can afford."[39] The urgency to slow or halt climate change has strengthened anthropologists' and others' predilection to praise agencies that disrupt, dismantle, and oppose, despite long-standing critique from feminist anthropologists of the dangers of equating agency with resistance. Collectively, academics working in this vein advocate for projects that seek insurgency rather than improvement, revolution rather than reform.

While courageous, such imaginings of potential futures largely have yet to consider what exactly we are to do with our existing infrastructure, knowledge, and persons besides critique or abandon them. Theorizing scholarly work that seeks to "enter and critically engage ecologies filled with people, things, and agencies that take for granted and perform persistent knowledge forms," Downey writes:

> We cannot expect interlocutors in technoscience to willingly inflect their own knowledge and put at risk persistent identities and commitments, especially when these appear in multiple, distributed ecologies of performance, such as in schools and at work. When sociotechnical critique includes travel, it recognizes itself as indicating alternate techniques, devices, infrastructures, and selves, and charges itself to critically examine their identities and personhood as they seek to facilitate learning.[40]

In this vein, Downey's proposal for reimagining engineering builds on the dominant image of problem solving but reframes it: technical mediation grounded in problem definition and solution with people who define and solve problems differently. This proposal, he argues, is more likely to be taken up by the people he seeks to reach, since it takes into account their own senses of self and profession.[41]

This book and the larger projects from which it emerged resonate with Downey's approach. To open up broader conversations about the entanglement of corporations in the pressing dilemmas we face, it has deeply explored how one crucial group of industry actors—engineers—experience and understand themselves, their work, and their participation in corporate forms. Putting that ethnographic material into conversation with social science leads me to propose that academics need new theories of how industrial harms come to be. These occur because of willful negligence, profit seeking, and deal cutting among elites, as existing literature points out. *But they also emerge from engineers' and other industry actors' attempts to reconcile competing domains of accountability as they craft themselves as "good" people and professionals.* This insight calls for understanding—and intervening in—engineering accountability as it is emergent in everyday practice, not just in dramatic moments of whistleblowing or in retrospective report writing. It also calls for a more subtle understanding of engineers' agencies outside of a binary between resistance and acquiescence to corporate forms. These can form the ground for proposing new images of accountability for engineers and the corporations they enact.

A LOYAL OPPOSITION?

While engineers' agencies are bound up in important ways with corporate forms, they are not determined by them. Engineers enact corporations through their everyday practices, leaving those larger entities themselves open to change, not just repetition and reification.[42] This book has traced how attention to social acceptance (and sometimes even social well-being) secured a crucial foothold in engineering decision making. A more profound challenge, of course, would be supporting engineers in imagining and helping bring about capitalism's "otherwises"—radically different forms of organizing entangled economic, political, social, and material life.[43] I suspect that the near impossibility of doing so from the position of corporate employment or even consulting is what led my interlocutors, who were the biggest questioners and skeptics, to seek respite in academia.

Seeing these limitations, some engineering educators envision future generations of engineering students abandoning corporate employment to work directly for social and environmental justice. While we ought to support and celebrate students who choose this path, it seems dangerous to imagine that we will fix the massive and complex problems posed by corporations simply by opposing them. Edwin T. Layton Jr., whose treatise on engineering and social responsibility (originally penned in 1968) continues to echo in contemporary debates, proposed an alternate path in his conception of a "loyal opposition." In this ideal, engineers (and their professional associations) would not abandon corporate work but would instead serve as internal sources of critique. He defined this vision in a 1983 review essay, in which he called for a "framework within which it will be legitimate for engineers to point out mistakes and threats to public safety, health, and welfare."[44] For him, the notion of loyal opposition would value dissent and constructive criticism rather than treating these capacities as treason.

The loyal opposition framework can provide a fertile starting point for reimagining engineers' work in corporate contexts. It opens up space for engineers to relate to corporate forms as both "we" and "they." It would highlight the multiple accountabilities embedded in engineers' work and give space for engineers to call out the gaps between them rather than directing most of their efforts to furiously attempting to reconcile them. Recognition of these multiple accountabilities would need to be based in a deep recognition that diverse judgments of the world exist and that harmonization of them is not always possible or desirable. This acknowledgment could seed more deliberative exchanges about how governments, industry, and citizens prioritize particular conceptual frameworks and courses of action as we imagine our interlinked social, environmental, and technological futures.

What kinds of engineers would we need to nurture to create such a loyal opposition? Many. To start: engineers who recognize how their decisions reinforce or challenge structures of power; engineers who can respectfully translate across different knowledge communities, whether that knowledge lives in disciplines or in the publics they seek to serve; and engineers who can reflect on the assumptions and aspirations guiding their decision making so that they can enter into difficult conversations with people who have different understandings of what the world is and what it ought to become.

Epilogue

Social science and humanities scholars can do better than "snapping at the world, as if the whole point of being and thinking is just to catch it in a lie."[1] Cultural critique is a vital and often undervalued scholarly agency, and critical reflection is a prerequisite to creating more just and caring worlds. Social science research unwaveringly shows that the technoscientific, corporate, and industrial forms we critique are borne out of global economic and political assemblages that are produced through the maneuverings of elites. But they are also partially brought into being through the deceptively mundane practices of people whose work brings corporate forms to life.

In this book I have argued that, to *understand* the accountability of technoscientific corporations, we must understand the agencies of the people who constitute them. In this epilogue, I sketch the contours of one way we might *alter* the accountability of technoscientific corporations by altering the agencies of the people who constitute them: critical participation in engineering education. The activities I analyze here emerged from much larger and collaborative attempts to create a space inside the Colorado School of Mines that was accountable to diverse scholarly identities, to the unique mission of the school, and to the aspirations of engineering faculty and students.[2] Much of this work involved other faculty at Mines, especially with engineers and social scientists affiliated with the Humanitarian Engineering program I directed with Juan Lucena. While I do not advocate for this approach as a universally desirable or the only mode of critical participation, I hope that making my efforts visible will inspire other critical reflection on our academic practices. Not all humanities or social science scholars aspire to enter into the worlds we critique, and forced participation is experienced as co-optation.[3] But each of us is already a participant in engineered systems as we live with the artifacts of engineering practice in our everyday lives. Those who teach engineers play a special role in shaping their positionality, subjectivity, and expertise.

Gary Downey's opening provocation in the journal *Engineering Studies* proposes "scalable scholarship" to avoid the "comforts of resolute pessimism and the pursuit of isolated critical virtuosity." Scalable scholarship takes seriously that which is otherwise dismissed as external to or derivative of our research: its relevance to and potential uptake by our students, our interlocutors, and other audiences.[4] Like others in the field, I seek, as Dean Nieusma proposes, "to produce scholarship that informs critical participation in engineering educational reform [and] to actively participate in such reform efforts and to reflexively analyze and refine that participation."[5]

Participation in education reform feels especially pressing in engineering, an early and prominent example of corporate influence over university curricula.[6] In my own case at Mines, my job talk—a critique of corporate social responsibility (CSR) in mining—took place in the then brand-new petroleum engineering building, where people filed into classrooms, lecture halls, labs, and study spaces underneath the full-color logos of oil and gas companies. Scholars argue that such influence has strategically blinded engineers to how "technical systems enable different structures of power."[7] Concerning the oil and gas industry in particular, science and technology studies (STS) scholar Sara Ann Wylie argues that companies and engineering universities promulgate a long history of promoting industrialization at the expense of rural people by "approaching natural gas extraction as a technical practice rather than as a political practice with social ramifications."[8] It is precisely the dominant image of engineering as a purely technical practice that my colleagues and I seek to interrupt in our teaching and pedagogical collaborations at Mines and elsewhere.

Our efforts in both program building and teaching make visible and disrupt a pervasive divide between the technical and the social that pervades engineering education and practice.[9] This divide has been a part of academic work since Enlightenment knowledge production systems organized knowledge into discrete disciplines, distinguishing the humanities from the sciences and technology from society. This system turned science into a factory, in which knowledge could be produced through a division of labor where each discipline has its own ontology, communication, and community.[10] This intellectual inheritance animates a long history of efforts to cordon off engineering knowledge from the humanities and social sciences and systematically devalue the latter. Those boundaries were reinforced over the past century in the United States, as practicing engineers and engineering educators called for scientizing engineering practice and knowledge in order to remain competitive in the Cold War.[11]

Those histories continue to reverberate in the undergraduate engineering curriculum encountered by students, as it artificially separates social from technical knowledge and systematically devalues the former.[12] In a large, longitudinal study of undergraduate engineering students at multiple institutions, Erin A. Cech demonstrated that engineering students' commitments to public welfare decreased the longer they were in school, what she identifies as a pervasive "culture of disengagement that defines public welfare concerns as tangential to what it means to practice engineering." She insightfully proposes that this culture of disengagement is rooted in the "ideology of depoliticization, which frames any 'non-technical' concerns such as public welfare as irrelevant to 'real' engineering work; the technical/social dualism, which devalues 'social' competencies such as those related to public welfare; and the meritocratic ideology, which frames existing social structures as fair and just."[13] This institutional context of engineering learning means that social responsibility can be taught in the most microethical of approaches, encompassing adherence to law, rather than a macroethical approach to critiquing the contribution of science, engineering, and technology to social and environmental (in)justice.[14]

No single definition of socially responsible engineering is taught at the undergraduate level, reflecting a broader lack of consensus in the profession as a whole, and very little research investigates what engineering students actually think socially responsible engineering is. Many retreat into small acts of personal responsibility and charity, while others hope that, by serving their employers and clients "faithfully and professionally, it will somehow all work out in the end."[15] Greg Rulifson and Angela R. Bielefeldt offer one of the few longitudinal, qualitative studies of how undergraduate engineering students actually understand social responsibility. For the students who stayed in engineering, Rulifson and Bielefeldt identified a convergence of themes that included "not harming environment, moving society forward, helping the impoverished, following the law, safety, serving company, helping people, behaving ethically, [and] serving clients/ end users."[16] Significantly, they found that over time themes such as moving society forward and helping the impoverished decreased in importance, whereas safety, serving the company, and serving clients/end users increased. This suggests that, much as in Cech's research, students came to view engineering as serving a smaller role in improving the lives of the marginalized and disadvantaged and to place greater importance on company loyalty as they progressed through the curriculum and internships. Moreover, Rulifson and Bielefeldt found that students with a high commitment to social responsibility often left engineering when they encountered unsupportive environments, decontextualized technical courses, and curricular difficulty.[17]

The intertwined ideology of depoliticization, technical/social dualism, and meritocratic ideology inform how engineering students come to think about their own accountabilities. I therefore aim to analyze and teach CSR as a sociotechnical field of practice inflected by structures of power, making visible and challenging the deeply divided institutional structures experienced by students and faculty. At the beginning of this intertwined ethnographic and educational research project on the intersection of engineering and CSR, I hoped that helping engineering students puzzle through the opportunities and limitations of CSR as a framework for engineering accountability would help mitigate the troubling and endemic culture of disengagement in undergraduate engineering education. I set out to use my ethnographic research to propose more robust notions of accountability for engineers from different spaces of the curriculum, from the social science elective I taught to their majors' core engineering courses.

Mines students are like other engineering undergraduates who have precious few opportunities to take courses in the liberal arts. Of the nearly 140 credit hours required for mining or petroleum engineering degrees at Mines, for example, a scant 15 are reserved for training in the humanities and social sciences: a first-year writing and environmental ethics course, a second-year global studies course, and three upper-division electives of a student's choosing. Nieusma underlines the potential power of STS-infused liberal education courses, even when few in number, "to provide the curricular, conceptual, and pedagogical frameworks to situate students' engineering coursework (i.e., the technical core) as well as their identities within a more expansive vision of engineering as occupation as well as of engineering in society more broadly."[18] I aspired for my Corporate Social Responsibility course to do this work for students at Mines, specifically by nurturing other kinds of agencies than students found in the rest of their technical coursework.

Since 2015 I have regularly taught Corporate Social Responsibility as an upper-division social science elective that facilitates a critical social science analysis of the concept and its use in practice. Through reading ethnography and industry reports, the students map out and critique the concepts and underlying assumptions that animate CSR as a field of practice. We identify an underlying politics of commensuration at work in the common practice of weighing trade-offs in engineering decision making.[19] Those trade-offs belie the harmony ideologies we encounter in most corporate publications, inspiring students to ask hard questions to wrestle with competing accountabilities. We analyze CSR as one among other frameworks for understanding the accountabilities of business to society. We explore how the CSR framing makes certain kinds of questions visible—or not—compared with other frameworks, such as government regulation, environmental justice, and sustainable development. We debate the merits and disadvantages of each framework for understanding corporate accountability. Students find that, for example, while the framework of environmental justice facilitates their bringing unequal distribution of benefits and harms to the fore of debates about business practice, CSR discourses steeped in the social license to operate more often led them to aspire to financial win-wins, such as those offered by the engineers in chapter 6.

Through these activities, we unearth the inherent politics of such terms as *stakeholder* and *social license to operate*. I invite students to see the power implicated in who gets to define legitimate stakeholders and or the "problem" of resource development engineers seek to solve. As much as possible, we try to widen our analysis beyond the operations phase privileged in very term *social license to operate* to consider the (in)justices involved in project planning.[20] This illustrates how the problem definition stages are where accountabilities that cannot be reconciled with profitability can be defined out of the "problem" of resource development. Hoping to upset the sociotechnical divide that would position CSR as a domain of anthropologists separate from engineers, I challenge students to redesign part of the project—such as by relocating wells, redesigning a pit, or creating new systems for safety management—based on the concerns of the publics they identify. When those concerns cannot be satisfactorily addressed through changes in engineering decision making, it places us squarely in discussions about *whether* industrial activity should take place, not just how it could be done responsibly.

To gain support for the new images of engineering and accountability I propose through the class, I enlist the support of engineers I have come to know during my research. These visits are some of the most meaningful parts of the class, especially by presenting students with a wider variety of potential engineering career trajectories than they are accustomed to imagining. A perennial favorite is Chuck Shultz, a Mines alum and oil and gas executive who has become one of our strongest supporters. Troubled that shortsighted engineering decisions have generated prominent conflicts, he has embarked on a mission to ensure that engineers can integrate social responsibility into their decision making, thus seeding greater accountability from the ground up. He has also inspired students to see the impact of leadership in setting a "tone at the top" that fosters social responsibility throughout the organization. He demystifies the corporate form by explaining the relationships among senior leadership, boards, and shareholder representatives. Speaking from his own experience, he shares how he has wrestled with difficult decisions that require him to balance accountabilities to shareholders, to employees, and to the people and places where they operate.

Our visitors consistently build up their proposals for new kinds of socially responsible engineers by intervening in existing professional ideals. Aaron's mission to change how his industry engaged the public, he made clear, required transforming the dominant images and practices of engineering inside of that industry (see chapters 1 and 6). In my class, he grounded his call for new kinds of engineering practices in the disciplinary identity he shared with the students, speaking with the *we* of engineers: "Engineers are problem solvers, but social relationships can't be solved, just managed. There's no formula. Engineers, we love to solve and delete problems, instead of working on the adaptive ones that don't go away because they are about relationships and context, and those change." Aaron invoked the powerful and pervasive identity of engineers as problem solvers but redefined the problems engineers were supposed to be solving to be the perplexing and ever-shifting ones of social acceptance and social responsibility.[21] For him, "managing" rather than solving those problems demanded characteristics that were simultaneously being advocated by some engineering education reformers, such as recognizing that other people may define problems differently, listening to those others to better understand their perspectives and concerns, and seeing engineering practice as sociotechnical rather than separable from political, economic, and social matters. I also hope to nurture those characteristics in my students through the course, making Aaron's visit a source of support from inside of an industry that holds substantial symbolic and financial capital on the Mines campus.

COLLABORATIONS WITH ENGINEERING PROFESSORS

The CSR elective course provides some opportunity to shape how students perceive what they encounter in the engineering core by presenting them with a social science critique of dominant modes of engineering and engineering education. But I was also eager to participate directly inside of engineering classes themselves, hoping to send the message that public accountability was central to the profession itself, not something that ought to be outsourced to social scientists.

In 2013, a year after my arrival to Mines, Carrie McClelland, a civil engineer with a background in community development who was teaching in the petroleum engineering department, invited me to lecture about the importance of stakeholder engagement in the required senior seminar she taught. Colorado was in the midst of the controversies accompanying the oil and gas boom (see chapter 1). To try to gain the trust of students I perceived to be skeptical, I began my lecture by making visible my personal ties to mining—a similarly vilified industry—and invoked the business-friendly "cost of conflict" to justify taking social and environmental performance seriously. I showed them different strategies for prioritizing stakeholders, leading them in an exercise in which they ranked stakeholders based on different criteria. My goal was for them to discover how stakeholder engagement was not an objective exercise but one steeped in politics, assumptions, and company policies.

We started by prioritizing criteria that privileged groups that were powerful squeaky wheels—those that wielded a considerable amount of influence to impact a company's ability to operate. I then invited them to consider other criteria, such as proximity, level of impact by the operation, and need for economic development. These criteria resulted in an entirely different ranking list, which provided the platform to propose that stakeholder engagement was a political activity inflected with the values of the person doing the ranking. I then gave them a few examples in which companies had chosen to explicitly exclude some potential stakeholders from the list in order to delegitimize them. McClelland's and my perception of the success of the experiment paved the way for more in-depth collaboration in the course over the next four years. We created assignments and role-playing exercises and brought in guest speakers from industry and my research project. By my last semester participating in the course, students felt comfortable enough even to propose social justice as one of the ranking criteria.

This work was bolstered by the National Science Foundation funding I received from the Cultivating Cultures for Ethical STEM program, which provided me the institutional legitimacy and funding to scale up those collaborations in her class and others. Nicole Smith and I convened a workshop in August 2016 that drew mining and petroleum engineering professors from Mines and the other participating institutions, plus CSR practitioners from the Denver area. We began interviewing engineers, creating productive synergies between the ethnographic research and our teaching activities. John, the engineer who worked on performance standard compliance for oil and gas projects around the world, worked with us to develop a semester-long project that invited students to step into the shoes of an engineer working on the controversial project in the South Pacific, charged with assuring its fidelity to the global performance standards for social and environmental responsibility. He also traveled to Golden, Colorado, to lecture to McClelland's class as well as to the campus as a whole.

Even though the senior seminar was required by the petroleum engineering degree program, and Carrie McClelland and I made efforts to use it to rethink dominant engineering practices, its status as the "professional skills" class led it being perceived as still separate from the technical core of the program. To begin working inside that core, we worked with Linda Battalora, a fellow professor in the department who held terminal degrees in both petroleum engineering and law. Crucially, Nicole Smith helped devise readings and activities about CSR for the required field session (figure E.1), an important rite of passage in which students spend a few weeks in oil-producing regions to tour production sites and learn firsthand from practicing engineers. Nicole Smith also attended one of the field sessions in California to lead those activities and insert questions about social responsibility into their site visits. In the following years, the field session faculty continued using the assignments we developed.

Linda Battalora also worked with us to institute two CSR-themed project-based learning exercises in two of the other courses she taught: Reservoir Fluid Properties and Mechanics of Petroleum Production. In each course, we hosted a mock "community meeting" similar to those that Aaron and others developed for communities undergoing unconventional energy development (chapter 4). Students created an exhibit that explained a dimension of one of their course themes—for example, oilfield

Figure E.1
Students visiting a production site during a field session for the Petroleum Engineering course, 2016. Photo courtesy Linda Battalora and Nicole Smith.

water—in a way that was meaningful to the people likely to attend such a meeting. We brought in students from my CSR class, other professors, and local research participants—including Art Biddle—to play the roles of various members of the public. We evaluated the students in their ability not just to present technical information but to listen and then understand and respond to expressed concerns in an appropriate manner. These efforts were significantly aided by the participation of Ray Priestley, a Mines alum and geological engineer who has taken an active role in visiting Humanitarian Engineering classes, developing assignments, and helping found an alumni interest group focused on social responsibility. I also engaged in more limited collaborations with other professors, including helping develop case studies and lectures for the Formation Damage and Stimulation course and Senior Design course, including working with John to create a lecture and assignment on risk assessment. Within a few years of the National Science Foundation project starting, petroleum engineering students encountered CSR in each year of their undergraduate training, inside of their "core" curriculum. The grant also allowed us to scale up the work to other courses at Mines, as well as at Virginia Tech, Marietta College, and the South Dakota School of Mines and Technology.[22]

ASSESSING CHANGES IN STUDENT KNOWLEDGE AND ATTITUDES

My colleagues and I were eager to gauge if and how our courses affect students' knowledge and attitudes surrounding the intersection of CSR and engineering. We have developed a variety of techniques for assessing changes in their knowledge, attitudes, and skills, including a survey instrument that we deploy in our courses at Mines, as well as in courses of our collaborators at Virginia Tech, Marietta College, and the South Dakota School of Mines and Technology. Students took the survey at the beginning of the course and again at the end, allowing us to compare their responses.[23] Between 2016 and 2020 we had compiled matched pre and post data for nearly one thousand students.[24]

One of our first findings was that our work in the petroleum engineering senior seminar led to students being more readily able to identify the social and environmental dimensions of a corporation's responsibilities to the public and its stakeholders. The students still struggled, however, to connect CSR directly to their professional activities as engineers.[25] We saw greater improvements in this area in the petroleum engineering field sessions.[26] In the first summer, for example, we observed a statistically significant change in student responses to the question, "To what extent do you think CSR will impact your work as an engineer?" At the start of the summer session, responses were roughly evenly split between "very probable" and "somewhat probable." By the end of the course, 100 percent of the students who participated in the field session with Nicole Smith and Linda Battalora responded with "very probable," along with about 70 percent of the students enrolled in the field sessions that used the same material but did not include a social scientist on the trip. That same summer, we observed statistically significant improvements in students recognizing that companies are accountable to "activists opposed to their industry" and "civil society organizations." In the field session attended by Nicole Smith, for example, just over half of the students began the session recognizing companies' accountabilities to activists. By the end, 90 percent did. We saw a smaller but still significant jump in the following year's field session, which used the same material but did not include a social scientist on the trip.

Our next studies broadened our scope to assess and compare student learning across the courses and universities. Overall, students were better able to define CSR and recognized accountabilities to a broader array of publics, such as activists and civil society organizations.[27] They also ended the courses expressing greater desires to work for socially responsible companies.[28] We found empirical evidence that women students at all of the institutions expressed greater desires for careers at socially responsible companies. We also found that students' opinions about social responsibility varied substantially by school, pointing to the significance of institutional culture. The mining engineering students at Virginia Tech, for example, consistently self-reported higher desires to work for socially responsible companies than did the engineering students anywhere else, including at the Colorado School of Mines.[29] While we do not know why Virginia Tech students' scores are so much higher than their peers' on the same question, we note that Virginia Tech actively promotes a culture of service among all of its engineering students.

Much to our chagrin, we did not find concrete evidence that a particular category of course—elective versus required, social science versus engineering—consistently enhanced students' attitudes about the intersection of engineering and social responsibility. Nor did we find a clear pattern in the timing of the course in their undergraduate curriculum, as we have evidence for statistically significant improvements for students in their first, second, third, and fourth years. We did find some evidence that specialized elective classes such as the CSR course I taught may be "preaching to the choir," as many educators suspect. Those students began with higher scores in response to the question asking them to rate their desire to work for socially responsible companies than any other of the Mines students in our sample.

The one course that consistently saw statistically significant improvement in students' knowledge and attitudes about social responsibility was a first-year seminar taught by petroleum engineering faculty at Marietta College. The students began the course with significantly lower expressed desires to work for socially responsible companies and weaker definitions of CSR than students at other universities but ended in the same range as their peers. Unlike the other courses, the seminar focuses on helping students make explicit connections between personal and professional responsibility, fostering what Nathan Canney and Angela Bielefeldt theorize as professional connectedness.[30] It is likely also significant that Marietta was the only liberal arts school included in our study, which could have a variety of effects on student self-selection into the program and the broader messages they are receiving about social responsibility.

To our surprise, the statistically significant improvements we observed in students' desires to work for socially responsible companies did not seem to correlate with those students coming to view CSR as an integrated, sociotechnical phenomenon. Our initial hope and hypothesis was that, if students could come to see CSR as sociotechnical rather than as simply social or technical in nature, they would be more likely to embrace it in their own career aspirations. Of the seven courses that demonstrated statistically significant changes in students' desires to work for a socially responsible company, only one also showed improvements in recognizing CSR as a sociotechnical practice. The other six showed that despite our best efforts, students were more likely to come to view CSR as the "old" CSR practices, such as volunteering, providing scholarships, and engaging in charity work, rather than the "new" CSR practices that made engineering itself more accountable to stakeholders. This finding may point to the difficulty of disrupting engineering students' tendency to engage in compartmentalization when imagining their future careers, viewing social responsibility as something important to them yet separate from the practice of engineering itself. We also, however, caution against judging the value of pedagogy by statistical significance alone. Students began many of the courses already recognizing CSR as a sociotechnical phenomenon, which would not result in gains that were statistically significant.

Our most disappointing finding was preliminary confirmation of a "social responsibility slump" as students progressed through their undergraduate curriculums. We were able to track four cohorts of students from year to year rather than just from the beginning to the end of a course. With few exceptions, these students ended their courses expressing greater desires to work for socially responsible companies, but they then began their *next* courses expressing *lower* desires than they had at the beginning of the previous course.[31]

One possible explanation is the social desirability bias in survey research, which shows that people answer sensitive questions with socially desirable rather than truthful answers. In our case, this means that students may have reported greater desires to work for a socially responsible company because they received signals during the semester that their professor held that view, and they were answering the survey questions in what they viewed as the "correct" way. We attempted to mitigate this bias by informing students in person and on the survey itself that their professor would not see their responses, but it is difficult to avoid this bias completely.

Aside from the social desirability bias, there are other possible reasons that students' attitudes may improve at the level of a class but lose ground as they advance through their undergraduate years. They may be experiencing the effects of the culture of disengagement documented by Cech. The technical/social dualism animated course flows at each of the universities and many students' perceptions of the value and difficulty of the "technical" courses inside of their majors and the "social" electives outside of them. Faculty perceived that students became immersed in their technical courses and could become a bit myopic when enamored of the new engineering concepts and industrial applications they are learning. Many of our students reacted strongly when they encountered CSR content in their technical courses, especially when those CSR concerns were raised as criticism of an industry they were aspiring to join.

The "culture of disengagement" argument focuses on students' experiences inside universities. For our own research, we cannot say that changes in student responses were solely due to their course experiences, since they all were taking other courses taught by other professors and absorbing information and perspectives from outside sources, such as news outlets, internships, and social networks. What we do suggest is that, in addition to the messages that engineering students receive inside universities and through the curriculum, students are powerfully affected by the messages they receive from practicing engineers and employers during internships.

Emily Sarver conducted precourse surveys of the students in the sophomore and senior mining engineering courses at Virginia Tech between 2011 and 2013, prior to our research project. She found that romanticism evident among the sophomores about their ability to use engineering to promote sustainability and social responsibility had tempered by the time they were seniors. For example, one sophomore student wrote, "I like that mining engineering is involved quite a bit with the business end of operations . . . and things like ensuring good social license is just as important as the engineering itself." Another sophomore wrote, "I like the direction that the industry is moving by putting such a focus on sustainability. I'm excited to be part of a generation of engineers that are more conscious of the environment than ever before and I think we can make a difference in an industry that has had such a dirty connotation for so long." In contrast, a senior wrote that the concepts in the Mine Reclamation and Environmental Management course "aren't really applicable to what most mining engineers are going to be doing in the industry." Another senior wrote that most of their "impressions of what to expect have come from professors and people I've met through internships. For the most part these people have shown a dislike for the environmental and regulatory requirements placed on them because they slow down production."[32] Working experiences may thus provide students an immediate sense of the multiple constraints that exist on engineering practice, in tension with social responsibility goals. This experience may prompt them to be more pragmatic or restrained in their social responsibility aspirations.

It may also be that seniors self-report lower desires to work for companies with positive reputations for social responsibility because they have already accepted job offers at companies with average or poor reputations in this regard and are calibrating their expectations to what they encountered on the job market. Moreover, the strength of the job market may make a difference in students' aspirations. In 2015, oil prices fell from near-record highs and production declined dramatically, reversing a boom in production and well-paying jobs that had originally attracted many students to study petroleum engineering. This market contraction meant that in 2015, 2016, and 2017 companies made fewer job offers and frequently rescinded offers that they did make. The lower scores for the earlier cohorts of petroleum engineering students may signal that they were willing to accept a job at any company, with less regard for their reputation, because they perceived that they had few options open to them.

In our ongoing and future research, we are keen to see if there is a relationship between students' career aspirations, their views of engineers' agency, and their views of corporations. Are students more likely to aspire to socially responsible careers if they believe that engineers can actively shape decision making that affects the well-being of people and the environment? If they view the activities undertaken under the banner of CSR as something that they will personally have to do as engineers? If they believe that CSR can actually improve industry practice instead of serving as public relations greenwash?

CONCLUSION

If CSR is a vehicle to extend the moral authority of corporations,[33] I have often worried that our collaborative teaching and institutional transformation efforts will facilitate that expansion. Many of our students go on to work in both large and small companies connected to mining and oil and gas activity, and I derive a great deal of satisfaction imagining how their engagement with critical social science will change how they do their engineering work—ideally to encompass the concerns of people who bear the greatest harms of industrial activity. Yet the more corporate personnel admit mistakes on a journey of "continuous improvement," the more they are able to incorporate stakeholder concerns into their outreach and operations, and the more they are able to extend the reach of corporate forms. This creates a potential catch-22: the more critical and self-reflexive my students become, perhaps the more they enable the expansion of the corporate forms employing them.

Doing nothing and retreating into social science research and teaching, disconnected from the teaching and practice of engineering, is also problematic, a form of purity politics that belies our personal and institutional imbrications with the mining and oil and gas industries in particular and with corporate forms in general.[34] I sympathize with my interlocutors' argument that we are all to some degree complicit in these industries—in different ways and to different degrees—given our dependence on them in our everyday lives. As problematic as the ethic of material provisioning is, it endures because it holds some truth. It also seems naive to pretend that our academic enterprises can be completely purified away from corporate interests, given the long history of corporate involvement in education in general, especially in engineering. We can make corporate influence visible and strategically work within, alongside, and against it, but we cannot ignore it. Finally, I see and hear a clear desire among engineers to learn from the research that social scientists have done.

The main concern of critical participation, then, becomes managing three risks outlined by Downey and Lucena: (1) co-optation, or the transformation of a project into "something indistinguishable from that which it studies"; (2) social engineering, or "presuming that one's expertise warrants the authority to legislate change through a research project"; and (3) rejection by the people with whom we seek to collaborate.[35] Our internal use of CSR concepts shored up institutional legitimacy for our projects of transforming engineering education but also risked cooptation by adopting the language of industry. To mitigate this, we presented CSR alongside other frameworks that questioned the universalization present in the cost-of-conflict market logic, posed different images of the obligations between companies and their publics, and raised the possibility that those publics defined "progress" or a "good life" differently than did engineering-trained corporate personnel.[36] To discourage social engineering and rejection, we took care to engage in "mutual critique."[37] I have opened up my research and teaching to critique by engineers by becoming a regular presenter at mining engineering, petroleum engineering, and engineering education conferences, and I have joined my collaborators and students in thinking through the strengths and limitations of ethnographic analysis. If we desire our engineering students to demonstrate more humility in accepting the limitations of their work and their larger disciplines, we must be willing to do the same.

It is in that spirit of humility that this epilogue has outlined some of the collaborative, interdisciplinary efforts at critical participation that co-evolved with the ethnographic research analyzed in the rest of the book. Quantitative survey measures can only tell part of the story of the impact of our teaching. I remain inspired by the many engineering students, professors, and practitioners who recognize—and sometimes relish—perplexity, using it as an opportunity to rethink their assumptions rather than denigrate people who hold different judgments of what the world is and what it ought to become.[38] I am buoyed by the many engineering students, professors, and practitioners who use those reflections to ask *whether* natural resource production should occur in particular places, in addition to how it can be done more responsibly. For those who seek transformation in the corporate forms that shape so many aspects of our everyday lives, we would do well to think with the people who enact them.

Notes

PROLOGUE

1. This has been a key insight especially of feminist scholars in anthropology and science and technology studies. For a short summary, see Carsten, Day, and Stafford, "Introduction."

2. Carsten, Day, and Stafford, "Introduction," 7.

3. Carsten, Day, and Stafford, "Introduction," 6.

4. Carsten, Day, and Stafford, "Introduction," 6.

5. Forsythe, *Studying Those Who Study Us*, 125.

6. Scheper-Hughes, "Ire in Ireland."

7. Forsythe, *Studying Those Who Study Us*, 123.

8. Rolston, *Mining Coal and Undermining Gender*.

9. Though the stresses of navigating corporate employment, temporary research positions, and university promotion and tenure systems are quite different, my institutional location created some resonances with Forsythe's observation that "it is one thing to write critically about events halfway around the world, or for a tenured professor to publish a critique of local power structures. It is quite another for an anthropologist in a corporation or on soft money *to call into question the practices of those who employ her and who may be in a position to affect her future ability to make a living.*" Forsythe, *Studying Those Who Study Us*, 123 (emphasis added).

10. Downey, "What Is Engineering Studies For?"

11. I am grateful to Beth Reddy for alerting me to this term and thinking with me about how it applies to those of us who work among engineers and applied scientists.

12. Shotwell, *Against Purity*, 107.

CHAPTER 1

1. Aaron's experience underscores Daniel Franks's observation that "change agents" inside corporations face challenges in being viewed with suspicion from both inside and outside but excel in drawing different worlds together. Franks, *Mountain Movers*.

2. For a critique of how the concept of stakeholders can privilege "consensus" that can reproduce social and environmental injustices, see chapter 6.

3. In this book I follow other social scientists in using *CSR* as an umbrella term that groups together a very diverse field of practices related to managing relationships between business and society. Companies and personnel label these activities with a variety of related yet distinct terms, such as *corporate responsibility*, *sustainability*, and *social performance*.

4. Winner, "Engineering Ethics and Political Imagination," 59.

5. Mitcham, "True Grand Challenge for Engineering."

6. Strathern, "New Accountabilities," 2.

7. Murphy, "Corporate Social Responsiveness."

8. Franks, *Mountain Movers*; Owen and Kemp, *Extractive Relations*.

9. Hawken, Lovins, and Lovins, *Natural Capitalism*; Porter and Kramer, "Creating Shared Value."

10. Wisnioski, *Engineers for Change*, 69.

11. Star and Griesemer, "Institutional Ecology, 'Translations' and Boundary Objects," 393.

12. Star, "This Is Not a Boundary Object," 604.

13. Blowfield and Frynas, "Setting New Agendas," 512.

14. Welker, "'Corporate Security Begins in the Community,'" 145; see also Fortun, *Advocacy after Bhopal*.

15. Lucena, "Engineers and Community"; Ottinger, *Refining Expertise*.

16. Appel, *Licit Life of Capitalism*; Barry, *Material Politics*.

17. Nixon, *Slow Violence*.

18. Ryan Cecil Jobson insightfully places coal-based industrialization, and its exploitation of wage labor, within a much longer history of plantation slavery turning human labor power into a thermodynamic reserve of potential energy. Jobson, "Dead Labor." See also Appel, *Licit Life of Capitalism*.

19. Kirsch, *Reverse Anthropology*; Jacka, *Alchemy in the Rain Forest*; Ottinger, *Refining Expertise*; Sawyer, *Crude Chronicles*.

20. Argenti and Knight, "Sun, Wind, and the Rebirth of Extractive Economies"; Howe, *Ecologics*.

21. Appel, *Licit Life of Capitalism*.

22. LeCain, *Mass Destruction*, 53.

23. LeCain, *Mass Destruction*, 209. As a case in point, in the Wyoming coal mines I first studied, crews had to move two to three tons of "overburden" (rock and dirt) to reach each ton of coal, leading some miners to joke that they would be more accurately described as "dirt haulers." Those ratios continue to increase as companies exhaust the deposits that are easier to access.

24. Priest, *Offshore Imperative*.

25. Boyer, "Infrastructure, Potential Energy, Revolution," 226.

26. Appadurai, "Mediants, Materiality, Normativity," 24. Quoted in Anand, Gupta, and Hannah, Introduction, 24.

27. For an overview of this tendency in academic research about mining and energy, see Smith and Tidwell, "Everyday Lives of Energy Transitions."

28. Berry, *Devoted to Nature*.

29. These life projects resonate with the "moral ambitions" of Mette M. High's interlocutors but were not as explicitly grounded in Christian religious beliefs. High, "Projects of Devotion."

30. Ballestero, *Future History of Water*, 4.

31. Mitcham suggests why this could be: "Engineers often believed that such loyalty was in the public interest because of their strong beliefs that capitalist corporations were themselves public benefactors." Mitcham, *Steps toward a Philosophy of Engineering*, 159.

32. Appel, *Licit Life of Capitalism*.

33. Appel, *Licit Life of Capitalism*, 5.

34. Following Marina Welker, Damani Partridge, and Rebecca Hardin, I refer to *corporate forms* rather than *corporations* in order to "productively shift away from default conceptualizations of corporations as solid, unified, self-knowing, and self-present actors that relentlessly maximize profits and externalize harm. Such an understanding of corporations, while appealing for its black-and-white guide to judgment, is divorced from history, geography, and actual corporate practice." Welker, Partridge, and Hardin, "Corporate Lives," S5.

35. Appel, *Licit Life of Capitalism*, 51; see 44 and 46 for visuals of the corporate archipelago.

36. Benson, "El Campo"; Müftüoglu et al., "Rethinking Access"; Rajak, *In Good Company*; Shever, "Engendering the Company"; Welker, *Enacting the Corporation*.

37. As Annelise Riles writes, "The fact that the corporation is 'one' from the vantage point of the state does not negate the way it might be 'many' from another point of view." Riles, "Too Big to Fail," 39.

38. Mol, *Body Multiple*.

39. Welker, *Enacting the Corporation*.

40. Williams, Muller, and Kilanski, "Gendered Organizations in the New Economy," 556.

41. As a case in point, AMAX, the mining company analyzed in chapter 3, also owned and operated the first large-scale coal mine in my hometown and my first research site. When AMAX tried to win public approval for the new mine in the 1970s, its personnel promised environmental stewardship and respectful treatment of miners. The mine was sold and purchased multiple times over the next four decades, until one of the companies declared bankruptcy, failed to make its required contributions to the miners' retirement accounts, and potentially left the state of Wyoming on the hook for remediation.

42. Kirsch, *Mining Capitalism*, 10; Appadurai, "Mediants, Materiality, Normativity," 16.

43. Laura Bear, Karen Ho, Anna Tsing, and Sylvia Yanagisako, "Gens: A Feminist Manifesto for the Study of Capitalism," Society for Cultural Anthropology, March 30, 2015, https://culanth.org/fieldsights/652-gens-a-feminist-manifesto-for-the-study-of-capitalism.

44. Ho, *Liquidated.*

45. Mitcham, *Steps toward a Philosophy of Engineering*, 22.

46. On infrastructures exceeding human lifetimes, see esp. Anand, Gupta, and Appel, *Promise of Infrastructure.*

47. Leydens and Lucena, *Engineering Justice*, 1.

48. Anand, Gupta, and Appel, *Promise of Infrastructure.*

49. Jacka, *Alchemy in the Rain Forest*; Jacka, "Anthropology of Mining," 22; Kirsch, *Reverse Anthropology*; Kirsch, *Mining Capitalism*; Sawyer, *Crude Chronicles*; Willow, *Understanding ExtrACTIVISM*; Wylie, *Fractivism.*

50. Folch, *Hydropolitics*; Barry, *Material Politics*; Harvey and Knox, *Roads*; Özden-Schilling, "Infrastructure of Markets"; Barnes, *Cultivating the Nile.*

51. Li, *Unearthing Conflict.*

52. Espig and de Rijke, "Unconventional Gas Developments."

53. Kneas, "Subsoil Abundance and Surface Absence"; Kneas, "From Dearth to El Dorado"; Kneas, "Emergence and Aftermath."

54. Hughes, *Energy without Conscience.*

55. Rajak, *In Good Company*, 236.

56. Welker, *Enacting the Corporation*, 65.

57. Rogers, *Depths of Russia*, 176.

58. Owen and Kemp, *Extractive Relations*, 223.

59. Owen and Kemp, *Extractive Relations*; Welker, *Enacting the Corporation*, 65.

60. Gilbert and Sklair, "Introduction," 4.

61. Ortner, "Resistance and the Problem of Ethnographic Refusal."

62. Ballestero, *Future History of Water*, 146.

63. Welker, *Enacting the Corporation*, 16.

64. Gilbert and Sklair, "Introduction"; Hughes, *Energy without Conscience*.

65. Hughes, *Energy without Conscience*, 152, 4.

66. Wisnioski, *Engineers for Change*, 4.

67. Mitchell, *Rule of Experts*; Scott, *Seeing like a State*.

68. Lucena, Schneider, and Leydens, "Engineering and Sustainable Community Development."

69. Boyer, "Infrastructure, Potential Energy, Revolution."

70. Harvey and Knox, *Roads*, 9.

71. Puig de la Bellacasa, "Ethical Doings in Naturecultures," 152.

72. Gilbert and Sklair, "Introduction."

73. High and Smith, "Introduction."

74. Ballestero, *Future History of Water*, 5.

75. High and Smith, "Introduction."

76. This question was posed eloquently by Dean Nieusma in his feedback on the manuscript for this book.

77. Indeed, one senior mining engineering professor told me that CSR was the "dessert" of the "main meal of mining," reinforcing the technical/social dualism critiqued in chapter 6. This trivialization of "the social" animates many technoscientific workplaces: even Forsythe recalls one of her interlocutors in artificial intelligence viewing her anthropological work as "frosting on the cake." Forsythe, *Studying Those Who Study Us,* 126

78. *Public* is a notoriously slippery term, perhaps most often used to refer to political collectives distinct from both markets and states. This book follows in the vein of work that emphasizes how publics congeal around infrastructure to make claims on governments and corporations. Infrastructures "show how publics are not just made through enunciatory communities, circulations of intention, text, and speech that produce disembodied spheres of deliberation and fantasies free of circulation. . . . Publics are also made by infrastructures that assemble collectives, constitute political subjects, and generate social aspirations." Anand, Gupta, and Appel, Introduction, 23. See also Barry, *Material Politics*.

79. Müftüoglu et al., "Rethinking Access," 250

80. Müftüoglu et al., "Rethinking Access," 257.

81. Forsythe, *Studying Those Who Study Us*, 125.

82. This research investigated the place of memorandums of understanding—put forward as a best practice in providing communities with more control over unconventional energy production—in Colorado fracking controversies.

83. Gusterson, *Nuclear Rites*, 115.

84. Memorable exceptions include Rajak, *In Good Company*; Rogers, *Depths of Russia*; and Welker, *Enacting the Corporation*.

85. Müftüoglu et al., "Rethinking Access," 257.

86. Boyer, "Thinking through the Anthropology of Experts," 43.

87. I conducted about half of the project's total 75 interviews, and anthropologist Nicole Smith conducted about half of the interviews while working as a postdoctoral scholar on National Science Foundation grant 1540298 that funded the research for this book. I follow ethnographic conventions in not identifying the interviewees by name, except for those in chapter 3 who also appear in the public records analyzed here and gave permission to be identified.

88. Mining engineering and petroleum engineering represent a relatively small proportion of engineering undergraduates in the US. The American Society for Engineering Education estimated that, in 2018, 245 bachelor's degrees in mining engineering were awarded across the country (0.2 percent of the 131,937 total for all engineering bachelor's degrees) and 2,118 were awarded (1.6 percent of total) in petroleum engineering. In contrast, the report estimated 31,936 bachelor's degrees in mechanical engineering. Joseph Roy, "Engineering by the Numbers," American Society for Engineering Education, July 2019, https://ira.asee .org/wp-content/uploads/2019/07/2018-Engineering-by-Numbers-Engineering-Statistics -UPDATED-15-July-2019.pdf.

89. Rajak, "Theatres of Virtue."

90. Conley and Williams, "Engage, Embed, and Embellish"; Müftüoglu et al., "Rethinking Access"; Rajak, *In Good Company*.

91. Downey, "What Is Engineering Studies For?"

92. Downey, *Machine in Me*, 199.

93. Kunda, *Engineering Culture*; Yarrow, *Architects*.

94. Yarrow, *Architects*.

95. Müftüoglu et al., "Rethinking Access," 2, 4.

96. Müftüoglu et al., "Rethinking Access," 4. See also Conley and Williams, "Engage, Embed, and Embellish"; and Dolan and Rajak, "Introduction."

97. Readers will also note that I refrain as much as possible from using the terminology of "extractive" industries. This is, first, to acknowledge the significant differences between the two industries and caution against social scientists inappropriately collapsing them together. Second, *extractive* is a term that indexes critique of these industries; it is not a term that any of my interlocutors used to describe their own work. Hughes's ethnography helps explain why the term offends petroleum geoscientists and engineers. He argues that they construct oil as a material that seems to desire to be on the planet's surface; rather than needing to be "extracted" like mined minerals that are forcibly hauled to the surface, oil is constructed in graphical representations of the subsurface and notions such as hydrocarbon

uplift as "naturally" seeking upward movement, such that the task of engineers and scientists is simply to liberate it so that oil "virtually produces itself." Hughes, *Energy without Conscience*, 65.

98. Rogers, "Materiality of the Corporation." See also Ferguson, *Global Shadows*; Kneas, "Subsoil Abundance and Surface Absence"; Kneas, "Emergence and Aftermath"; and Richardson and Weszkalnys, "Introduction."

99. Welker, *Enacting the Corporation*.

CHAPTER 2

1. Strathern, "New Accountabilities," 4.

2. Trnka and Trundle, "Introduction," 5. See also Laidlaw, "Ethics"; and Kelty, "Responsibility." For an influential critique of those neoliberal forms of personhood see Rose, *Politics of Life Itself*.

3. Li, *Unearthing Conflict*. In contrast, Trnka and Trundle argue that responsibility is predicated on recognition of the Other. Trnka and Trundle, "Introduction," 18. It is much more common for engineers and those who study them to use the term *social responsibility*. This preference may reflect the diffuse imagination of public welfare that animates most professional engineering societies' ethical codes.

4. Downey, "What Is Engineering Studies For?"

5. Latour, "Why Has Critique Run Out of Steam?"

6. Johnson, "Rethinking the Social Responsibilities of Engineers," 90.

7. Johnson, "Rethinking the Social Responsibilities of Engineers," 96–97.

8. In practice, however, ethics codes played a strikingly minor role in how the engineers I came to know thought about and practiced accountability to their coworkers and the public: not one of them mentioned referencing a professional code for guidance on a tough decision. Instead, they referenced corporate policies, legal statutes, and standards.

9. Drawing on fine-grained ethnography of the Denver headquarters of Newmont Corporation—the world's largest gold mining company—and its Batu Hijau copper-gold mine in Indonesia, Marina Welker argues that profit maximization forms a "large and loose target, an imprecise orienting device rather than a clear roadmap prescribing a fixed route for corporate managers and staff to follow." She interprets invocations of the business case as "claim-making devices that people deploy in particular contexts in order to justify or support particular courses of action." Welker, *Enacting the Corporation*, 17.

10. For a more in-depth analysis of Joe's experiences, see Smith et al., "Plea for Enhancing Engineering Ethics."

11. Mitcham, *Steps toward a Philosophy of Engineering*. For more in-depth engagement with this idea, see chapter 7 of this book.

12. Penelope Harvey and Hannah Knox argue that engineers can point to standards as a way of detaching personal responsibility for their actions, which likely holds true for the other formal accountability schemes. Drawing on their research with civil engineers working on a major infrastructure project in Peru, they argue, "The political formation of the regulatory frame points us to the ways in which standards are devices that allow engineer to control without personal responsibility, without having to acknowledge their involvement, and certainly without having to confront the issue of what gets left out of the picture once the regulations are applied." Harvey and Knox, "Virtuous Detachments," 67.

13. Mitcham argues that the practical effect of the loyalty clause in engineering ethics codes was to "undermine independence" and promote a "self-imposed tutelage to capitalist corporate employers." Mitcham, *Steps toward a Philosophy of Engineering*, 159.

14. Gary Downey concisely summarizes Layton's original influential argument and critiques of it: "The story of what Edwin Layton called the *Revolt of the Engineers* is told by him as a victory of engineering leaders associated with industry, by Noble as a victory of corporate capital, and by Meiksins as the product of a temporary alliance between elite reformers and rank and file engineers who accepted locations within companies but sought to increase income." Downey, "Low Cost, Mass Use," 298.

15. Through the committee, Jackson developed "a wide ranging Canon of Ethics and spent the rest of his time as chairman persuading various regional and national engineering organizations to adopt it. His embrace of the view of engineers as businessmen, coupled with the pro-business orientation of the codes of ethics that emerged during the Progressive Era, lends credence to concerns that corporate interests seep into engineering ethics education even outside of recognized CSR discourses." Smith and Lucena, "Socially Responsible Engineering," 664.

16. Mitcham, *Steps toward a Philosophy of Engineering*.

17. Downey, "Low Cost, Mass Use." See also Layton, *Revolt of the Engineers*; Meiksins, "'Revolt of the Engineers' Reconsidered"; Meiksins and Smith, "Why American Engineers Aren't Unionized"; and Noble, *America by Design*.

18. Wisnioski, *Engineers for Change*, 67.

19. Mitcham, "Historico-ethical Perspective on Engineering Education"; Tang and Nieusma, "Contextualizing the Code"; Wisnioski, *Engineers for Change*.

20. See IEEE, "Homepage," https://www.ieee.org (accessed January 1, 2021).

21. Wisnioski, *Engineers for Change*, 68. It is telling that none of my interlocutors pointed to their professional ethics codes as shaping how they reasoned through dilemmas. A few referenced external standards in the form of World Bank/International Finance Corporation performance standards or ISO (International Organization for Standardization) certifications but found that they had to be justified by positioning them in relation to their company's own CSR policies.

22. Mitcham, *Steps toward a Philosophy of Engineering*.

23. Williamson, "Small-Scale Technology for the Developing World"; Wisnioski, *Engineers for Change*.

24. Mody, *The Squares*.

25. Lucena, Schneider, and Leydens, "Engineering and Sustainable Community Development"; Leydens and Lucena, *Engineering Justice*.

26. Herkert, "Future Directions in Engineering Ethics Research."

27. Downey, "What Is Engineering Studies For?"

28. Smith, "Ethics of Material Provisioning."

29. Appel, "Conclusion," 181.

30. Appel, "Conclusion," 181.

31. Dahlgren, "Digging Deeper," 174.

32. The "humans" Kim invokes are actually a privileged group, as not everyone uses Amazon or drives a car.

33. LeCain, *Mass Destruction*, 191–193.

34. Huber, *Lifeblood*, 71, 72.

35. See American Petroleum Institute, "News, Policy, and Issues," https://www.api.org/news-policy-and-issues/top-industry-policy-issues (accessed July 12, 2020).

36. Smith and Tidwell, "Everyday Lives of Energy Transitions."

37. LeCain, *Mass Destruction*, 186; Wisnioski, *Engineers for Change*, 11–12.

38. Chapman, "Multinatural Resources; Huber, *Lifeblood*, 90; Hughes, *Energy without Conscience*, 309.

39. Hochschild, *Strangers in Their Own Land*.

40. Downey, "Low Cost, Mass Use."

41. Brueckner and Eabrasu, "Pinning Down the Social License to Operate (SLO)"; Owen and Kemp, *Extractive Relations*.

42. Boutilier and Thomson, *Social License*, 8. For an earlier iteration, see Thomson and Boutilier, "Social Licence to Operate."

43. Owen and Kemp, "Social Licence and Mining," 30.

44. Brueckner and Eabrasu, "Pinning Down the Social License to Operate (SLO)."

45. Delborne, Kokotovich, and Lunshof, "Social License and Synthetic Biology."

46. Gehman, Lefsrud, and Fast, "Social License to Operate."

47. Thomson and Boutilier, "Social Licence to Operate." Stuart Kirsch places it a year earlier in the paper and pulp industry, which was seeking to avoid new government regulations. Kirsch, *Mining Capitalism*, 209.

48. These consultants wielded great power to transform industry practice but also came under fire for their "greater fealty to their employers than to the indigenous people these processes were intended to protect." Kirsch, *Mining Capitalism*, 212.

49. Owen and Kemp, *Extractive Relations*, 30.

50. For a critique of Shell's CSR reporting, see Livesey and Kearins, "Transparent and Caring Corporations?"

51. Knight, "Profits and principles," 20.

52. Davis and Franks, "Costs of Company-Community Conflict in the Extractive Sector"; Franks et al., "Conflict Translates Environmental and Social Risk into Business Costs."

53. On the limitations of Ruggie's framework, see Kirsch, *Mining Capitalism*, 211.

54. Brueckner and Eabrasu, "Pinning Down the Social License to Operate (SLO)"; Prno and Slocombe, "Exploring the Origins of 'Social License to Operate' in the Mining Sector"; Thomson and Boutilier, "Social Licence to Operate."

55. Boutilier and Thomson, *Social License*.

56. Owen and Kemp, "Social Licence and Mining"; Welker, "'Corporate Security Begins in the Community.'"

57. Gardner, *Discordant Development*; Rajak, *In Good Company*.

58. Kirsch, *Mining Capitalism*; Ottinger, *Refining Expertise*; Rajak, *In Good Company*.

59. Delborne, Kokotovich, and Lunshof, "Social License and Synthetic Biology," 4.

60. Kirsch, *Mining Capitalism*, 209.

61. Owen and Kemp, "Social Licence and Mining," 34.

62. Katy Gardner argues that rumors are a way for people to make sense of what is going on around them, including by expressing and perpetuating fear, which offers us a way of seeing the world as our narrators see it. Gardner, *Discordant Development*, 194, 222.

63. Smith et al., "Industry–University Partnerships."

64. This was a common analogy among my most senior interlocutors, who argued that safety first moved from being a cost of doing business to being part of corporate missions, followed by environmental protection, and then a variety of "social" concerns of sustainable development and community acceptance. The lack of stability for a term for this domain of practice signals its ongoing contestation and the lack of fit with distinct "social" goals, such as social risk management and sustainable development.

65. Nader, "Controlling Processes."

66. Smith and Helfgott, "Flexibility or Exploitation?," 20. See also Cloud, "Corporate Social Responsibility as Oxymoron"; and Eagleton, *Ideology*.

67. His invocation of shareholders supports Kirsch's argument that the SLO is a technique to assure shareholders that a company can manage risk. Kirsch, *Mining Capitalism*, 209.

68. Smith, "From Corporate Social Responsibility to Creating Shared Value."

69. Welker, *Enacting the Corporation*, 65. See also Owen and Kemp, *Extractive Relations*, 226.

CHAPTER 3

1. Most of the mining industry personnel I met had moved away aspirations for "sustainable mining" to instead focus on mining's contributions to sustainable development. The most stringent critique I heard was from June, a mining engineer who candidly said, "The other thing that always gets to me as an engineer is, we call this sustainable development. To me, that's a joke. These are not sustainable industries. These are extractive industries. When the extractive industry leaves, there's going to be an impact on the community no matter what you do. All you can do is try and do something [good]." Academics have been far more critical of the proposal that mining can be sustainable, including by arguing that the term sustainable mining is a "corporate oxymoron." Kirsch, "Sustainable Mining."

2. Frynas, "Political Instability and Business"; Watts, "Blood Oil."

3. Rolston, *Mining Coal and Undermining Gender*.

4. The chapter draws from both Stan Dempsey's and Art Biddle's personal archives—which included letters, speeches, notes, newspaper clippings, and public relations materials—as well as multiple interviews with each. I also worked extensively with Biddle in the CSR educational activities I developed at the Colorado School of Mines, described in the prologue and epilogue; he participated as a judge for student projects, presented his experiences in my CSR course, and attended social events bringing together Mines students and alumni. To contextualize Dempsey's and Biddle's work, I also interviewed engineers and applied scientists who worked on these projects as AMAX employees, as consultants, and as state agency officials and made trips to visit the museums and archives in both Gunnison and Crested Butte, Colorado, where I also discussed the mining controversy with residents and historians there. I name Dempsey and Biddle because they took on public-facing roles in the debates and gave me permission to do so, but I do not use the real names of the other people mentioned in this chapter, as they did not take on public roles in the debates. This chapter's focus on two individuals fits within broader calls for "life writing" as a complement to ethnography, which otherwise focuses on social patterns instead of particular individuals. Carsten, Day, and Stafford, "Introduction."

5. Appel, "Walls and White Elephants"; Gilberthorpe and Banks, "Development on Whose Terms?"; Jacka, "Anthropology of Mining"; Kirsch, *Mining Capitalism*; Li, *Unearthing Conflict*; Rajak, *In Good Company*.

6. Barandiarán, *Science and Environment in Chile*.

7. Li, "Documenting Accountability," 218.

8. See chapter 6 for a more in-depth consideration of engineering pragmatism.

9. Franks, *Mountain Movers*.

10. Indeed, Dempsey recognized that the language of cost effectiveness would help to win over engineers to his more progressive views of environmental management and social acceptance,

and he invoked the specter of costly delays forty years before the landmark "Cost of Conflict" study described in chapter 2. One retelling of the Experiment in Ecology stated: "When working with AMAX managers and engineers, Dempsey does not talk about saving the environment just for the environment's sake. He seeks rather to frame the problems and solutions in terms of cost effectiveness. He tries to show the manager how the environmentally correct way will pay off in the long run even if it costs more in the immediate time frame. Dempsey's chief argument is that doing the right kind of environmental homework can prevent years of costly delay." Cahn, *Footprints on the Planet*, 76.

11. Kelty, *Participant*, 36.

12. Increased attention to the environment in the 1970s followed on the heels of the 1950s and 1960s uptick in federal workplace safety regulations.

13. Hoffman, *From Heresy to Dogma*, 12.

14. Cumming, "Black Gold, White Power."

15. Jenkins, *Decade of Nightmares*.

16. Kaiser, *How the Hippies Saved Physics*; Kaiser and McCray, *Groovy Science*.

17. Mody, *The Squares*.

18. Gottlieb, *Forcing the Spring*, 204; Wellock, *Critical Masses*, 112.

19. Hoffman, *From Heresy to Dogma*, 12.

20. Barandiarán, *Science and Environment in Chile*, 9

21. Barandiarán, *Science and Environment in Chile*, 24.

22. Mathews and Barnes, "Prognosis."

23. Barandiarán, *Science and Environment in Chile*; Li, "Documenting Accountability"; McNeil, *Combating Mountaintop Removal*; Ottinger, *Refining Expertise*. Moreover, an in-depth comparison of environmental impact projections and actual water quality performance at hard rock mines revealed that the projections systematically underestimated potential pollution. Kuipers et al., "Comparison of Predicted and Actual Water Quality at Hardrock Mines."

24. Barandiarán, *Science and Environment in Chile*, 37.

25. The baseline studies were done for Henderson mine, and the first full environmental impact statement was for the Belle Ayr coal mine in Gillette, Wyoming. Dempsey argued that, whereas other the companies seeking to develop mines in the booming Powder River Basin hedged their bets on overturning NEPA and its requirements for impact assessment, he took the impending regulatory changes as actuality and began drafting the EIA before NEPA was actually passed in to law. This allowed AMAX to get Belle Ayr up and running before the 1974 Supreme Court injunction that temporarily halted coal development in the basin, beating the other mines to coal production by at least five years. See Rolston, *Mining Coal and Undermining Gender*, chap. 2.

26. AMAX was the product of a merger, in this case, between American Metal Co. Ltd. and Climax Molybdenum Company in 1957.

27. AMAX made the case for the continued and expanded moly production by invoking the ethic of material provisioning in advertisements, a company-wide magazine, the Mt Emmons project's *Moly Newsletter*, speeches, and published documents. A 1981 article about environmental permitting by Biddle and colleagues explained that moly can "improve strengths, hardenability, weldability, corrosion-resistance, abrasion-resistance, and heat-resistance in steels for building spacecraft and aircraft parts, Arctic pipeline, building structures, solar panels, railroad rails, automobiles and machinery. Moly is also used in chemicals and lubricants, including engine oils and grease, pigments in paints and plastic, fertilizers, seed treatments and foliar sprays." Biddle, Livermore, and Poe, "AMAX Inc and the Colorado Joint Review Process."

28. See, e.g., *Journal of Metals*, "Henderson Mine/Mill/Concentrator"; Paxton, "Experiment in Ecology"; and *Engineering and Mining Journal*, "Openness, Cooperation."

29. Government agencies included the US Forest Service; Bureau of Land Management; US Fish and Wildlife Service; Colorado Game, Fish and Parks Department; State Bureau of Mines; and Colorado Water Pollution Control Commission.

30. In 1999, the company received an Environmental Awareness Award from the US Forest Service for the design, engineering, construction, and implementation of a fifteen-mile overland conveyor system that replaced the original rail line.

31. *Journal of Metals*, "Henderson Mine/Mill/Concentrator," 12.

32. The total cost of the original construction was $500 million, making it the largest private investment in Colorado history at the time.

33. *Journal of Metals*, "Henderson Mine/Mill/Concentrator," 12.

34. MacGregor was a Scottish metallurgical engineer and vocal opponent of organized labor. After retiring from AMAX, he became infamous for his role in Margaret Thatcher's 1980s industrial restructuring of the United Kingdom. As chairman of British Steel, he was ruthless in closing plants and trimming the payroll from 166,000 employees to 71,000. He took a similar approach in his leadership of the UK National Coal Board, which contributed to the 1984–1985 mineworkers strike and the prolonged decline of Britain's coal industry.

35. Dempsey attributed this quote to Pierre Gousseland, who replaced MacGregor as AMAX chairman in 1977. Dempsey made this connection explicit in a 1980 speech to the Business and Industry Advisory Committee of the OECD in Paris: "AMAX Environmental Services Inc. made a special effort to develop new project skills in the 1970s because of AMAX's fast expansion, and because our biggest problems were with regulatory delay and politically motivated permit denials." Stan Dempsey and Ken Paulsen speech notes, "AMAX and Environmental Impact Assessments," Business and Industry Advisory Committee of the OECD, Paris, September 26, 1980, 2, Stan Dempsey archives.

36. Dempsey, speech notes for Caitlin talk, 1, Stan Dempsey archives.

37. Stan Dempsey to MacGregor, Donahue, and Sawyer, February 20, 1974, 1, Stan Dempsey archives.

38. Paxton, "Experiment in Ecology," 8.

39. Jean Bettie Willard to Jean Langenheim, March 1, 1986, in "Bettie Willard, Alpine Ecologist," *Ecological Society of America's History and Records*, https://esa.org/history/willard-bettie/.

40. In his 1977 speech to the National Association of Manufacturers, Dempsey described the ecologists with whom he and other AMAX personnel worked "responsible citizen activists." Describing the Minnamax project, he said, "We made a conscious decision to avoid endless dialogue with citizen activists who were clearly dedicated to opposition to any mine, regardless of its environmental acceptability." Dempsey, NAM speech, Washington, DC, (March 25, 1977), 6, 9, Stan Dempsey archives. A consultant observed this in practice at both the Minnamax and Crested Butte projects, recalling that the AMAX personnel "were trying to work with the environmentalists and separate sort of the reasonable ones from the unreasonable ones, and I suspect the number on the reasonable side of the bench was pretty low." Social scientists have noted the tendency of corporations to elect to work with "light green" groups that advocate market-based solutions to problems rather than "dark green" groups that take a more critical stance. See Kirsch, "Sustainable Mining"; Rolston, "Turning Protesters into Monitors"; Welker, "'Corporate Security Begins in the Community.'"

41. *Engineering and Mining Journal*, "Openness, Cooperation," 2. Describing critiques of mining as being "emotional" was clearly a rhetorical strategy of dismissal.

42. Hoffman, *From Heresy to Dogma*, 13.

43. The name *Minnamax* was intended by mine proponents to blend the names of the company and the state, performing harmony between them.

44. AMAX of Minnesota, Inc., "Phase I a Six-Year Study: Minnamax Copper-Nickel Project, 1974–1979, a Period of Testing and Evaluation" (1980), 9, Art Biddle archives.

45. In 1974 the Minnesota Environmental Quality Council ruled that AMAX did not need to produce an EIS for sinking the test shaft on the site, a decision that was affirmed by the Minnesota Supreme Court in 1975 after environmental groups challenged the original decision. The Minnesota Pollution Control Agency granted permits for the shaft in July 1975.

46. These efforts happened alongside the state of Minnesota's regional multimillion-dollar copper-nickel study. For greater detail on those studies and their lasting influence for an engineer who worked for the Minnesota state government, see chapter 5.

47. Drawing on critical studies of audit culture, Marina Welker calls these "rituals of verification" that uphold the company's authority. Welker, *Enacting the Corporation*, 10.

48. Jim Blubaugh, "Sierra, Waltons End Opposition to Copper Test Shaft," *Duluth News-Tribune*, March 18, 1975. See also Art Biddle speech notes, National Conference on the Management of Energy-Environment Conflict, Queenstown, Maryland, May 20–23, 1980, 4, Art Biddle archives.

49. Stan Dempsey, speech to the International Iron and Steel Institute meeting, Colorado Springs, Colorado, October 2, 1978, 4, Stan Dempsey archives.

50. Dempsey criticized, however, the institutional separation of environmental and social performance away from operations, as he believed that they needed to be integrated throughout the company or otherwise risked being marginalized.

51. Phadke, "Green Energy Futures."

52. The other key public official was the town planner, Myles Rademan, who had been persuaded to move to Colorado from New York City by "a slide show of the Rockies and really good grass" offered by a University of Colorado law professor seeking to recruit idealistic young lawyers. Dawn Belloise, "Profile: Myles Rademan," *Crested Butte News*, June 25, 2015, http://crestedbuttenews.com/2015/06/profile-myles-rademan/.

53. Crested Butte Mountain Heritage Museum, Crested Butte, Colorado.

54. Gladwin would go onto a long academic and consulting career, culminating in retiring from the University of Michigan Ross School of Business as Professor Emeritus of Sustainable Enterprise and Professor Emeritus of Strategy.

55. Thomas Gladwin, "Constructive Corporate Management of Environmental Conflict," Keystone conference, Keystone, Colorado, September 7, 1977, 17, Art Biddle archives.

56. John Towers, "Opening Remarks, Keystone Conference," Keystone, Colorado, September 7, 1977, 4, Art Biddle archives.

57. Cahn, *Footprints on the Planet*, 79.

58. The "triple bottom line" concept has gained widespread use inside of industry and beyond it, after its first articulation by John Elkington in a 1994 article in the *California Management Review* and then in his formulation of the 3P framework of "planet, people, and profit" in Shell's watershed 1998 corporate responsibility report "Profits and Principles." For a more critical take on triple bottom line, see Norman and MacDonald, "Getting to the Bottom of 'Triple Bottom Line.'"

59. This is an early example of what Stuart Kirsch calls the "new politics of time," though the Mt Emmons case pushes the "newness" back to the 1970s instead of 1990s and 2000s. Kirsch, *Mining Capitalism*.

60. These included the US Forest Service (the lead agency on the EIS), Colorado Department of Local Affairs, Colorado Department of Health, Colorado State Historical Society, Bureau of Land Management, US Environmental Protection Agency, and US Army Corps of Engineers.

61. Biddle, Livermore, and Poe, "AMAX Inc and the Colorado Joint Review Process," 1.

62. Mike Rock, speech notes for the annual meeting of the American Society of Landscape Architects, Denver, Colorado, November 21–24, 1980, 5, Art Biddle archives.

63. Other, shorter-lived concerns included the electric power line and whether the area could build a local power plant to serve the mine.

64. "Kay Ferrin transcript," CJRP meeting, Gunnison, Colorado, March 13, 1981, 1–7, Art Biddle archives.

65. "Kay Ferrin transcript," 1. Biddle introduced each of the AMAX speakers by noting their specific enjoyment of the outdoors. In so doing, he crafted a public face of the company as one that shared environmental values with the people of the area. See chapter 4 for a more extended discussion of how corporate actors "played the scales" of the corporate person to attribute an environmental ethic to companies.

66. *Gunnison County Times,* July 16, 1979.

67. "Kay Ferrin transcript," 5. Ferrin began by making fun of highly technical mining jargon by saying, "The milling operation will be a typical sulfide mineral hydro-metallurgical beneficiation project—see, I could say that, which means that we grind that rock up into real small pieces in large grinding mills, ball mills, and then float it in a detergent solution, like big washtubs, and separate the concentrates from the ore at that point and then the tailing will be deposited immediately below the mill." "Kay Ferrin transcript," 3.

68. "Kay Ferrin transcript," 6.

69. "Kay Ferrin transcript," 13–14.

70. "Second Question and Answer Period—March 13 CJRP meeting," Gunnison, Colorado, 1981, 1–3, Art Biddle archives.

71. Art Biddle, "August Status Report," Mt Emmons External Affairs Status Reports, July/August 1979, 6. Mt. Emmons Mining Project collection.

72. Dempsey described this happening at Henderson as well: "Mutual respect developed, and the project turned out better because all of us were forced to think through the reasons for our actions." Dempsey, NAM speech, 7.

73. Stan Dempsey speech notes, ESI, Environmental Managers Workshop, Denver, Colorado, November 9, 1978, 6, Stan Dempsey archives.

74. David Isaacson, "AMAX, Inc.," Encyclopedia.com, https://www.encyclopedia.com/books /politics-and-business-magazines/amax-inc (accessed January 1, 2021).

75. Art Biddle internal memo, "Questions and Answers, Proposed 1981 Community Affairs Budget," 3, Art Biddle archives. Peter Benson and Stuart Kirsch would likely characterize these efforts to promote the inevitability of the mine as engendering a politics of resignation. Benson and Kirsch, "Capitalism and the Politics of Resignation."

76. Biddle, "Questions and Answers," 8, Art Biddle archives.

77. As reported by AMAX in the Spring 1982 *Moly News,* vol. 5 no. (1).

78. For Biddle, this adoption of the EIS by the US Forest Service represented an achievement, as it signified the agency's approval of their work and support of the project. For those critical of the project, this was evidence of a too cozy relationship between AMAX and the agency.

79. David E. Leindorf and Stan Smock, Gunnison County commissioners, and County Planning Commission, to Jimmy Wilkins, US Forest Service, reproduced in the Final Mt. Emmons Mining Project EIS, January 20, 1982, 92. Mt. Emmons Mining Project collection.

80. Thomas S. Cox, mayor of the Town of Crested Butte, to Jimmy Wilkins, US Forest Service, reproduced in the Final Mt. Emmons Mining Project EIS, 109–110.

81. Leindorf et al. to Wilkins, Final Mt. Emmons Mining Project EIS, 92. Mt. Emmons Mining Project collection.

82. US Forest Service, "Purpose and Need," reproduced in the Final Mt. Emmons Mining Project EIS, 2.

83. US Forest Service, "Public Involvement," reproduced in the Final Mt. Emmons Mining Project EIS, 27.

84. Dempsey, NAM speech, 7.

85. Dempsey, NAM speech, 12.

86. Dempsey, NAM speech, 15.

87. Dempsey and Paulsen, "AMAX and Environmental Impact Assessments," 4.

88. Dempsey and Paulsen, "AMAX and Environmental Impact Assessments," 1. Here Paulsen evokes what Silvio O. Funtowicz and Jerome R. Ravetz would call "postnormal technologies," characterized by high controversy and experts competing with dueling studies that cannot resolve debates that are fundamentally about values. Funtowicz and Ravetz, "Science for the Post-normal Age."

89. Ferguson, *Anti-politics Machine*.

90. Sue Wilson, TJ Glauthier, and Ellen Remmer, "Trip Report: Interviews on Colorado Joint Review Process," January 19–22, 1982, 2. Mt. Emmons Mining Project collection.

91. Ottinger, *Refining Expertise*. See also Kelty, *Participant*.

92. Conley and Williams, "Engage, Embed, and Embellish," 13.

93. Dempsey, NAM speech, 1977, 13.

94. Art Biddle, "The Mount Emmons Projects and the Colorado Joint Review Process: One Approach for Gaining Community Acceptance through Better Coordination of Local, State, and Federal Permitting Requirements," June 17, 1986, 2. Art Biddle archive.

95. Harvey and Knox, "Virtuous Detachments in Engineering Practice." In chapter 6 I analyze pragmatism in detail.

CHAPTER 4

1. US Energy Information Administration, "State Profile and Energy Estimates: Colorado Profile Analysis," March 19, 2020, https://www.eia.gov/state/analysis.php?sid=CO.

2. Haggerty et al., "Geographies of Impact and the Impacts of Geography."

3. This slippage between the legal fiction of a composite, intangible corporate person and the material persons charged with enacting it originally stemmed from Christian theology before entering Tudor legal doctrine and the English parliamentary state. See Shever, "Engendering the Company," 29; and Welker, *Enacting the Corporation*, 3.

4. Kelty, *Participant*, 19.

5. Kelty, *Participant*, 19.

6. Organizational psychologists have called this a "we-feeling," which is often an explicit goal of human resource management designed to promote cooperation among workers and identification with a firm while, in the eyes of critics, suppressing internal unrest. Kelty, *Participant*, 110.

7. The literature on the mutual constitution of agency and structure is too extensive to adequately summarize here, but see Ahearn, "Language and Agency."

8. Latour, *Reassembling the Social*; Callon, "Some Elements of a Sociology of Translation."

9. Boas, "Methods of Ethnology," 316.

10. Bourdieu, *Outline of a Theory of Practice*; Certeau, *Practice of Everyday Life*; Ortner, "Theory in Anthropology since the Sixties"; Giddens, *Central Problems in Social Theory*.

11. Butler, *Gender Trouble*; Foucault, *History of Sexuality*.

12. Abu-Lughod, "Can There Be a Feminist Ethnography?"; Ortner, "Theory in Anthropology since the Sixties"; Ortner, "Resistance and the Problem of Ethnographic Refusal"; Mahmood, *Politics of Piety*.

13. Mahmood, *Politics of Piety*, 14.

14. Mahmood, *Politics of Piety*, 34.

15. Layton, *Revolt of the Engineers*, 198; Meiksins, "'Revolt of the Engineers' Reconsidered"; Meiksins and Smith, "Why American Engineers Aren't Unionized"; Noble, *America by Design*.

16. Kunda, *Engineering Culture*, 15.

17. Kunda, *Engineering Culture*, 218.

18. Edwards, *Contested Terrain*, 148.

19. Mauss, "Category of the Human Mind."

20. Mcintosh, "Personhood, Self, and Individual."

21. Laidlaw, "Ethics." Not all connotations of *soul* are so positive, as Mette M. High's ethnography of the "cosmoeconomies" of gold in Mongolia makes clear. See High, *Fear and Fortune*.

22. Johnson, "Rethinking the Social Responsibilities of Engineers," 95.

23. As analyzed in greater detail in chapter 5, engineers' status as employees and clients has undermined their status as professionals.

24. Downey, "What Is Engineering Studies For?," 60–61.

25. Roland Marchand traces how Progressive Era advertisements for the Chicago meatpacker Swift, for example, portrayed the company as a farm wife feeding chickens, and those for AT&T depicted the company as a female telephone operator and male lineman, setting a

precedent that other companies would emulate for years to come. Marchand, *Creating the Corporate Soul*.

26. Müftüoglu et al., "Rethinking Access," 7.

27. Kunda, *Engineering Culture*, 159.

28. Dinah Rajak argues that the field of corporate social responsibility in general is pervaded by proselytizing, in which executives and practitioners strive to convince others that they can be financially profitable by being socially responsible. Rajak, *In Good Company*.

29. Chapter 5 takes up the position of consultants in detail, especially their double bind as dependent on corporations for contracts yet being asked to be impartial assessors of those corporations' activities.

30. In this, they seemed very conscious that the close alignment would be dismissed as insincere. A senior environmental engineer insisted that her commitment to the environment she had just shared with me was "not the company line." She continued, "I don't want to sound like a corporate billboard, but we do want to do the right thing because all these people who live in Wyoming live here because they want to. They enjoy the outdoors in some regards more than others. We don't want to trash our environment because this is where we play and work."

31. For a similar point in relation to the ritualized confession of mistakes and commitments to learning from them, see Müftüoglu et al., "Rethinking Access," 7.

32. David Graeber summarizes Marilyn Strathern's notion of the partible person by explaining, "People have all sorts of potential identities, which most of the time exist only as a set of hidden possibilities. What happens in any given social situation is that another person fixes on one of these and thus 'makes it visible.' . . . Other possibilities, for the moment, remain invisible." Graeber, *Toward an Anthropological Theory of Value*, 39–40.

33. Mcintosh, "Personhood, Self, and Individual," 4583.

34. Mcintosh, "Personhood, Self, and Individual," 4583.

35. Geertz, "Thick Description"; Lamb, "Making and Unmaking of Persons."

36. Strathern, *Gender of the Gift*, 13.

37. Here I happily join others in exploring more relational theories of personhood in a US context. See Buch, "Senses of Care."

38. Welker, *Enacting the Corporation*.

39. Yarrow, *Architects*, 165.

40. Li, *Unearthing Conflict*; Smith and Smith, "Engineering and the Politics of Commensuration."

41. Unconventional energy production articulates with underground geology to produce an economic imperative: the quickly declining production curves of "tight oil" feed the need for more and more wells to keep supplies constant. An environmental engineer with

experience in mining, offshore oil, and onshore oil and gas production explained the uniqueness of unconventional gas production clearly, describing it as

> a bit more of an ad hoc type of business. You put a well in, you produce, the production declines, you frack it, you put another well in. It's constantly expanding, versus we're going to put an offshore platform that's going to pump for fifty years. That's very different, right? We're struggling with how to manage that kind of business. And then, odds are you've got an operator, another operator over here, and another operator over here, and another operator over here. You're doing an impact assessment on a well that you know is going to be producing for two years. At the same time, you're putting fifteen others in. The other guy is putting a hundred others in, and the other guy is putting two hundred others in. You're looking at it piece by piece.

He wished that regulations were different to require more regional studies of this kind of extractive activity rather than considering them on a well-by-well basis.

42. Welker and Wood, "Shareholder Activism and Alienation."

43. On the role of religious convictions in the moral ambitions of oil executives, see High, "Projects of Devotion." On the religious underpinnings of corporate social responsibility work in the South African mining industry, see Rajak, *In Good Company*.

44. Compare Gary's story about leaving industry and Scott's founding of a nonprofit function inside his consulting firm for engineers trying to set up professional spaces not determined by profit imperatives; see chapter 5.

45. Lucena, Schneider, and Leydens, "Engineering and Sustainable Community Development"; Leydens and Lucena, *Engineering Justice*.

46. Research suggests that this is also true for engineering students, as women with greater desires for social responsibility leave undergraduate engineering programs. Rulifson and Bielefeldt, "Motivations to Leave Engineering."

47. Williams, Muller, and Kilanski. "Gendered Organizations in the New Economy"; Williams, Kilanski, and Muller, "Corporate Diversity Programs and Gender Inequality"; Miller, "Frontier Masculinity in the Oil Industry."

48. Rolston, *Mining Coal and Undermining Gender*.

49. Of course, her "socially accountable" pit design did not take into account the desires of the local people who wished for there to be no mine at all.

50. The narrative June told provides an example of what John Owen identifies as a key feature of the SLO framework: the fear of losing access. Owen, "Social License and the Fear of Mineras Interruptus."

51. Kusserow, "De-homogenizing American Individualism"; Mcintosh, "Personhood, Self, and Individual"; Smith, "From Dividual and Individual Selves to Porous Subjects."

52. Colby and Sullivan, "Ethics Teaching in Undergraduate Engineering Education"; Mitcham and Wang, "From Engineering Ethics to Engineering Politics"; Zhu, "Engineering Ethics Education"; Zhu and Jesiek, "Pragmatic Approach to Ethical Decision-Making in Engineering Practice."

CHAPTER 5

1. There is significant diversity within the world of professional consulting for engineers and applied scientists. Some worked for large, well-established firms with an international reputation for particular specialties, while others ran one-person independent shops. While some were lifelong consultants, others moved into that work late in life to ease into retirement after successful corporate careers. Still others went into consulting after having been fired; as one seasoned consultant wryly observed, "Everybody that loses their job then calls themselves a consultant." The youngest consultants I met sought out consulting because it fit well with their self-described "entrepreneurial" spirit and desire to forge their own career path. A few even moved from consulting work to full-time corporate careers in order to "make a difference." Though it is difficult to pin down the exact percentage of technical professionals who work in these industries as consultants, mining engineering faculty I interviewed estimated that about 10 percent of their graduates eventually worked as consultants, but only after gaining experience and their professional engineering license through working for an operating company. Consulting work was more common for civil, environmental, and geological engineers. About 30 percent of the people interviewed for this book worked as consultants. Of those, about equal numbers worked for large, well-established firms or their own one-person shops.

2. Engineers who worked full-time for corporations also narrated the importance of "choosing" the right employer that aligned with their values. Their stories were strikingly similar, beginning with an acknowledgment that they were skeptical of unscrupulous corporate actors before choosing to work for a company well regarded for its ethical reputation. Applying to particular companies, performing in interviews in ways that demonstrate belonging to them, and accepting job offers are important dimensions of engineers' agencies. But these agencies were conditioned by structural factors such as interview processes, formal systems for evaluating knowledge, and social capital in the form of professional networks. For example, the companies that my students most admired also tended to have the highest requirements for minimum grade point averages, meaning that students with lower GPAs had fewer opportunities to "choose" their employers.

3. Barandiarán, *Science and Environment in Chile*, 171.

4. Strathern, "Cutting the Network"; Yarrow et al., *Detachment*.

5. Cross, "Detachment as a Corporate Ethic," 35.

6. Rajak, *In Good Company*.

7. Gardner, *Discordant Development*; Gardner et al., "Elusive Partnerships."

8. Shever, "Engendering the Company."

9. Appel, "Walls and White Elephants," 445.

10. Cross, "Detachment as a Corporate Ethic," 36.

11. Harvey and Knox, "Virtuous Detachments," 171.

12. Stein, *Work, Sleep, Repeat*; Dougherty, "Boom Times for Technocrats?"

13. Welker, *Enacting the Corporation.*

14. Barley and Kunda, *Gurus, Hired Guns, and Warm Bodies*, 304. On how the new career models characterizing increasingly precarious employment in the oil and gas industry have gender bias baked into them, see Williams, Muller, and Kilanski, "Gendered Organizations in the New Economy."

15. Stein, *Work, Sleep, Repeat*, 89.

16. Barley and Kunda, *Gurus, Hired Guns, and Warm Bodies*; Kunda, *Engineering Culture*; Meiksins and Whalley, *Putting Work in Its Place*. In addition to further theorizing this form of engineers' agency and efforts to cultivate professional autonomy, this insight adds a novel perspective to theories of the corporation. My interlocutors bring our attention to the question of *when* a corporation is enacted. While corporations are enacted by various personnel, from community relations officers to consulting engineers, those people are not always enacting those forms. Put another way, the parts or nodes of the set of relationships that make up a corporation are themselves not always enacting the larger corporate form. The employees and consultants called to enact corporations did so in some social contexts but not in others, and sometimes wholeheartedly, sometimes partially, sometimes with a wink, and sometimes not at all. They managed the playing of scales between their person and the corporate person (see chapter 4) moment by moment, turn by turn.

17. Goldman, "Why We Need a Philosophy of Engineering," 166.

18. Layton, *Revolt of the Engineers*, 5.

19. For summaries of that literature, see Davis, "Professional Autonomy"; and Meiksins and Watson, "Professional Autonomy and Organizational Constraint."

20. Barley and Kunda, *Gurus, Hired Guns, and Warm Bodies*, 291.

21. Barley and Kunda, *Gurus, Hired Guns, and Warm Bodies*, 289.

22. Dougherty, "Boom Times for Technocrats?"

23. Barley and Kunda, *Gurus, Hired Guns, and Warm Bodies*, 289.

24. Li, "Engineering Responsibility," 64.

25. Li, "Engineering Responsibility," 64. Framing debates about the question of *how* versus *whether* to mine was a key strategy developed by AMAX to manage public controversy surrounding their projects, as described in chapter 2.

26. Li, "Engineering Responsibility," 64.

27. Here Peter's narrative of his career echoes the performance of skepticism analyzed in the chapter 4.

28. Peter wished that companies would be less concerned with avoiding reputational risk and openly share their lessons learned with the wider academic and industry communities. "The companies have some really good data, but they don't want to share it because they're afraid that it's going to highlight something that's going to cause them more problems. . . .

It's like, nothing is perfect, but this is really a glass half-full story. And we should be telling it, but I can't get them to tell it because people [outside of industry] look at it as glass half empty." Stan Dempsey, the AMAX executive profiled in chapter 3, felt a similar frustration. He stopped conducting the internal environmental audits he developed at AMAX because he was worried they could set up the environmentalists' cases against the company if they were to fall into their hands.

29. See chapter 7, note 9, on the desire to help.

30. While Scott appreciated being part of a group that shared his political sensibilities, he also saw the drawbacks to hiring "a bunch of identical people" and specifically tried to hire different kinds of people so that his firm could bring a wider variety of perspectives to their projects.

31. Harvey and Knox, *Roads*, 10.

32. Gary traced this desire to a specific moment in his professional development. While contracting for major multinationals, he jumped at the chance to travel professionally in Eastern Europe and Russia when the Berlin Wall came down in 1990. He found himself fielding such questions as, "How do you teach your kids capitalism?" He recalled that when he tried to use garage sales as a way of answering that question, he experienced culture shock, realizing that his interlocutors did not have cars—and therefore no private garages—or the privilege of having old yet functional stuff to discard. Gary experienced culture shock a second time on that trip when he eventually made his way to the multinational's headquarters in Moscow, where he and the other "experts" on the project convened in a "Lufthansa hotel . . . drinking French wine imported through California." From that moment on, he found it impossible to detach issues of mineral development from broader concerns of culture and political economy.

33. Stein, *Work, Sleep, Repeat*.

34. PolyMet Mining Corp., "Polymet Strengthens Permitting Expertise Groundwater Monitoring Requirements Satisfied," Newsfile, September 11, 2012, https://www.newsfilecorp .com/release/2842/Polymet-Strengthens-Permitting-Expertise-Groundwater-Monitoring -Requirements-Satisfied.

35. John Meyers, "Safety of PolyMet Tailings Basin Dams Is Point of Contention in Permit Process," Twin Cities Pioneer Press, August 28, 2017, https://www.twincities.com/2017 /08/28/safety-of-polymet-tailings-basin-dams-is-point-of-contention-in-permit-process/; emphasis added.

36. PolyMet Mining Company, "Tailings Basin Stability and Environmental Protections" brochure. March 30, 2017. https?//www.polymetmining.com.

37. Phadke, "Green Energy Futures," 163.

38. PolyMet Mining Corp., "Polymet Strengthens Permitting Expertise Groundwater Monitoring Requirements Satisfied," Newsfile, September 11, 2012, https://www.newsfilecorp .com/release/2842/Polymet-Strengthens-Permitting-Expertise-Groundwater-Monitoring

-Requirements-Satisfied. Not all of the consultants and their firms are local, however, as the case of June profiled in this chapter makes clear.

39. Alder, *Engineering the Revolution*.

40. Downey, "PDS."

41. Meiksins and Watson, "Professional Autonomy and Organizational Constraint," 578.

42. Quoted in Leydens and Lucena, *Engineering Justice*, 4.

43. Zussman, *Mechanics of the Middle Class*; Miller, "Professionals in Bureaucracy."

44. Dougherty, "Boom Times for Technocrats?," 452. See also Barandiarán, *Science and Environment in Chile*.

45. Harvey and Knox, *Roads*, 196.

CHAPTER 6

1. Coloradans for Responsible Energy Development, "About Us," https://www.cred.org/about-cred-coloradans-for-responsible-energy-development/ (January 1, 2021).

2. Coloradans for Responsible Energy Development, "Scientists Agree: Fracking Doesn't Harm Our Water," https://www.cred.org/scientists-fracking-doesnt-harm-water/ (accessed January 1, 2021).

3. Rajak, "Theatres of Virtue"; Müftüoglu et al., "Rethinking Access"; Welker, *Enacting the Corporation*.

4. Many advocates for engineers celebrate their pragmatic agency in building the infrastructures, products, and processes that constitute our world. Former BP executive and engineer John Browne, for example, opens his 2019 book by praising engineers as "best known for their practical impact; while others talk and pontificate, they are out in the world, influencing and shaping it. If you look around, you will see a world made richer, freer, and less violent by engineering." Browne, *Make, Think, Imagine*, 1–2.

5. Minteer, "Pragmatism, Piety, and Environmental Ethics."

6. Reddy, "Measuring like an Engineer"; Riley, "Engineering and Social Justice."

7. Discourses of the social license to operate are firmly grounded in this particular pragmatism—the social license, after all, values good community relations as good for business (chapter 2).

8. Mitcham, *Steps toward a Philosophy of Engineering*, 2.

9. Appel, "To Critique or Not to Critique?," 32.

10. Ottinger, *Refining Expertise*, 133.

11. Faulkner, "'Nuts and Bolts and People.'"

12. Cech, "Culture of Disengagement in Engineering Education?"

13. Auld, Bernstein, and Cashore, "New Corporate Social Responsibility."

14. Philanthropy remained a powerful image of social responsibility for my interlocutors as well. Professionals in community relations termed such activities "strategic investments" and viewed them as most effective when they supported the overall mission of their companies. A petroleum engineer who had achieved significant influence by building up a successful private company with her family pointed to the strategic importance of philanthropy for creating an educated and healthy workforce. Speaking with the *we* of the oil and gas industry in Denver, she called this work "proactive":

> We work with a lot of philanthropic organizations, and the Denver oil and gas community is incredibly generous with community service and projects, raising funds for education, for the arts and for the health community. . . . And I think that ties in with being socially responsible. That's a different avenue of here you've got the side of, "Let's protect the people and environment where we're working." This is actually, "Let's enhance our community so that we've got kids going to college and we've got good health programs and good health research programs going on in our community." So it's a lot more than just that reactive or preventive method. It's actually trends are now toward a proactive environment.

15. Downey, "What Is Engineering Studies For?," 56.

16. Downey, "Are Engineers Losing Control of Technology?"

17. Lucena, Schneider, and Leydens, "Engineering and Sustainable Community Development," 125.

18. Here Aaron acknowledges a difference between Hannah Appel calls lack and loss: whereas corporations often position their CSR activities as addressing a lack in the communities closest to them, she argues they are often mediating a loss (of environmental quality, of traditional livelihoods, etc.) that they themselves created. Appel, *Licit Life of Capitalism*.

19. Li, *Unearthing Conflict*. Social science research shows that these economic benefits can create or heighten inequalities, as mineral owners receive substantial money in the form of royalties, whereas surface owners receive much smaller payments for surface land disturbance, and other nearby residents may receive nothing at all outside of the funds directed to local governments. For a summary, see Jacquet, "Review of Risks to Communities from Shale Energy Development."

20. Nader, "Controlling Processes." Science and technology studies scholars would also rightly point out that shared infrastructure and design for community acceptance are grounded in a form of "techno-optimism" that proposes technological solutions for problems that are fundamentally political in nature.

21. Collier and Ireland, "Shared-Use Mining Infrastructure," 20.

22. In her research on water and mine conflicts in Peru, Fabiana Li shows how water use does not always map onto legal designations, meaning that mining companies can take legal shelter while worsening water quality by arguing that it was not suitable for human consumption to begin with. Li, *Unearthing Conflict*.

23. Sellwood, "Peru's Fight for Millions in Tax Revenue from Cerro Verde Mine," Oxfam, September 20, 2017, https://politicsofpoverty.oxfamamerica.org/perus-fight-for-millions-in-tax-revenue-from-cerro-verde-mine/. For the more positive case studies of the plant, see International Council on Mining and Metals, "Shared Water, Shared Responsibility, Shared Action: Cerro Verde, Peru," March 21, 2017, https://www.icmm.com/en-gb/case-studies/cerro-verde; and Christopher Connell, "This Peruvian Mine Produces Clean Water for Arequipa," Share America, December 21, 2016, https://share.america.gov/peruvian-copper-mine-also-produces-clean-water/.

24. Munoz and Burnham, "Subcontracting as Corporate Social Responsibility"; Li, *Unearthing Conflict*.

25. Gardner, *Discordant Development*; Welker, *Enacting the Corporation*.

26. Love and Garwood, "Electrifying Transitions."

27. Kroepsch, "New Rig on the Block"; Kroepsch, "Horizontal Drilling."

28. Kroepsch, "Horizontal Drilling," 470; emphasis added.

29. Ottinger, *Refining Expertise*; Li, *Unearthing Conflict*.

30. Virtual frack centers, in which engineers can monitor wells from a distance, may be replacing this on-the-ground learning about the context of oil and gas development.

31. Appel, "Walls and White Elephants"; Rajak, *In Good Company*; Welker, *Enacting the Corporation*.

32. Smith, "From Corporate Social Responsibility to Creating Shared Value."

33. Li, "Engineering Responsibility"; Kirsch, *Reverse Anthropology*.

34. Horowitz and Watts, *Grassroots Environmental Governance*.

35. International Finance Corporation, "Performance Standards," https://www.ifc.org/wps/wcm/connect/Topics_Ext_Content/IFC_External_Corporate_Site/Sustainability-At-IFC/Policies-Standards/Performance-Standards (accessed July 17, 2021).

36. Hommels, *Unbuilding Cities*.

37. John R. Owen and Deanna Kemp identify this temporal relegation as one of the key limitations of community relations work in mining. Owen and Kemp, "Social Licence and Mining."

38. Callon, Lascoumes, and Barthe, *Acting in an Uncertain World*; Hébert, "Chronicle of a Disaster Foretold."

39. This stands in contrast with other engineers who evinced what Matthew Wisnioski, in *Engineers for Change*, calls an "ideology of technological change" that attributed change to the technology itself rather than to those who produced it, thus absolving engineers of responsibility for its effects. The most prominent example of the more "autonomous" technological change was how petroleum and other engineers told the story of the fracking revolution itself. With vertical drilling, the story goes, engineers had to spread out multiple wells across

a set area of space to access the oil and gas below. With the advent of horizontal drilling and hydraulic fracturing, in contrast, engineers could concentrate fewer multiple wells radiating out in multiple directions on a single pad, thus reducing their overall footprint on the surface. For a critique of how this concentration of wells on a single pad generates more harms for people who live closest to enlarged well pads, see Kroepsch, "New Rig on the Block"; and Kroepsch, "Horizontal Drilling."

40. Harvey and Knox, "Virtuous Detachments in Engineering Practice."

41. Lucena, Schneider, and Leydens, "Engineering and Sustainable Community Development."

42. See, e.g., Jalbert, Kinchy, and Perry, "Civil Society Research and Marcellus Shale Natural Gas Development"; Kinchy, "Citizen Science and Democracy"; Ottinger, *Refining Expertise*; and Wylie, *Fractivism*. Benjamin's experiences, however (see chapter 5), show how transformative more community-based research can be for how engineers think about the purpose of their work.

43. The growing power of premining referenda may be providing more space for people to say no to both mining companies and the national governments who facilitate their activity. Kirsch, *Mining Capitalism*.

CHAPTER 7

1. Kelty, *Participant*, 19.

2. Ballestero, *Future History of Water*, 53, citing Nader, "Controlling Processes."

3. Ballestero, *Future History of Water*.

4. Downey, "PDS."

5. Direct opportunities to listen to critics did not always result in increased empathy, however, as Emma's story of her exchange with the disgruntled landowner reveals in chapter 4.

6. Cech, "Culture of Disengagement in Engineering Education?"; Rulifson and Bielefeldt, "Motivations to Leave Engineering."

7. Trevelyan, *Making of an Expert Engineer*.

8. National Society of Professional Engineers, "NSPE Code of Ethics for Engineers," https://www .nspe.org/resources/ethics/code-ethics. My thanks to Dean Nieusma for raising this point.

9. Donna Riley argues that the "desire to help" is a persistent engineering mindset that can serve social justice ends only if implemented in a nonpaternalistic manner. As critics of philanthropy argue, "help" can often shore up the good feelings of the helper without fundamentally transforming the relationships of power that create injustice in the first place. Riley, *Engineering and Social Justice*, 39.

10. Owen and Kemp, "Social Licence and Mining."

11. Biersack, "Reimagining Political Ecology," 14. See also Haraway, *Modest Witness@Second Millennium*; Latour, *We Have Never Been Modern*; and Strathern, *After Nature*.

12. Richardson and Weszkalnys, "Introduction," 7. See also Bakker and Bridge, "Material Worlds?"; Barnes, *Cultivating the Nile*; and Kneas, "Emergence and Aftermath."

13. Ferry and Limbert, *Timely Assets*; Weszkalnys, "Anticipating Oil"; Metze, "Framing the Future of Fracking."

14. This appreciation for nature as more than resource was sometimes prompted by crisis, such as when John read anthropological work on kinship and land to understand indigenous resistance to the oil and gas facilities his company was building.

15. For analysis and history of the conflict, see Rolston "Turning Protesters into Monitors."

16. Discourses of environmental stewardship are direct responses to criticisms of industry practice. The notion of environmental stewardship, cast in the mold of settler colonialism, entails particular relations of responsibility that legitimize industrial management of nature. Suzana Sawyer argues, "The authority of corporate capital today is related in important ways to historical practices of imagining, representing and purifying 'natural' landscapes . . . the way Arco imagined the terrain of its operations significantly affected its rights, responsibilities and legitimacy to explore for and exploit petroleum in Ecuador." Sawyer, *Crude Chronicles*, 103. Geographer Gavin Bridge argues that mining corporations developed discourses of environmental stewardship to mediate the tensions among accumulation, production, and environmental protection in order to legitimize their continued operation. He traces how officials have co-opted the language of nongovernmental organizations in their public policy statements by embedding ecological concerns within their business practice. Bridge, "Excavating Nature," 222–223, 227.

17. Bowker, "Sustainable Knowledge Infrastructures," 211.

18. Barry, *Material Politics*.

19. Kelty, *Participant*; Ottinger, *Refining Expertise*.

20. Kelty, *Participant*, 25–26.

21. Kelty, *Participant*, 30.

22. Dahlgren, *Digging Deeper*.

23. Owen and Kemp, *Extractive Relations*, 34.

24. Delborne, Kokotovich, and Lunshof, "Social License and Synthetic Biology."

25. Smith, "Boom to Bust, Ashes to (Coal) Dust."

26. Cronon, "Trouble with Wilderness."

27. Chakrabarty, "Climate of History."

28. A representative of one of Canada's First Nations critiqued the industry's paternalism and reliance on the ethics of material provisioning during a major mining conference. He said, "I hear it often in these kinds of events. There's a real bemoaning of the industry saying, 'If only people understood us.' It kind of sounds like the dad talking to the kids. 'If you guys

only understood what I did for you, how hard I worked for you to put food on the tables, minerals in your phones, then you would appreciate me.'"

29. Kelty, *Participant*, 56.

30. Kelty, *Participant*, 83.

31. Appel, "To Critique or Not to Critique?," 32.

32. Gilbert and Sklair, "Introduction"; Carrier, *After the Crisis*.

33. Appel, "To Critique or Not to Critique?," 32.

34. Bowker, "Sustainable Knowledge Infrastructures," 211.

35. Karwat, "Self-Reflection for Activist Engineering," 37

36. Bowker, "Sustainable Knowledge Infrastructures," 205.

37. Mitcham, *Steps toward a Philosophy of Engineering*, 259.

38. Downey, "What Is Engineering Studies For?"

39. Boyer, "Infrastructure, Potential Energy, Revolution," 231.

40. Downey, "Critical Participation," 14.

41. Elsewhere Downey writes, "Changes for the future always have to begin with what is positioned as given in the present. Even fundamental challenges to the hegemony of dominant practices have to address the question of fit with dominant practices." Downey, "Engineering Cultures Syllabus as Formation Narrative," 428.

42. Welker, *Enacting the Corporation*, 32.

43. Appel, *Licit Life of Capitalism*; Ballestero, *Future History of Water*.

44. Layton, *Revolt of the Engineers*.

EPILOGUE

1. Berlant and Stewart, *Hundreds*, 42.

2. Many of those broader efforts—which included hosting a campus lecture series, creating new university organizations, and collaborating with engineering faculty in their classes— echo the strategies that Jon Leydens and Juan Lucena propose for integrating social justice into engineering problems, assignments, projects, courses, programs, and universities. Leydens and Lucena, *Engineering Justice*.

3. Kelty, *Participant*.

4. Downey, "What Is Engineering Studies For?," 74, 57.

5. Nieusma, "Analyzing Context by Design," 417.

6. Downey, "Low Cost, Mass Use"; Noble, *America by Design*; Seely, "Research, Engineering, and Science in American Engineering Colleges."

7.   Wylie, *Fractivism*, 289; see also Leydens and Lucena, *Engineering Justice*.

8.   Wylie, *Fractivism*, 127.

9.   Faulkner, "'Nuts and Bolts and People'"; Cech, "Culture of Disengagement in Engineering Education?"

10. Bowker, "Sustainable Knowledge Infrastructures."

11. Lucena, *Defending the Nation*.

12. There are notable exceptions to this general rule, such as Olin College, which has no departments or tenured faculty in an attempt to foster collaboration and focus engineering on the needs of people in the real world.

13. Cech, "Culture of Disengagement in Engineering Education?," 45.

14. Herkert, "Future Directions in Engineering Ethics Research."

15. Catalano, "Engineering Ethics."

16. Rulifson and Bielefeldt, "Evolution of Students' Varied Conceptualizations," 939.

17. Rulifson and Bielefeldt, "Motivations to Leave Engineering."

18. Nieusma, "Conducting the Instrumentalists," 160. See also Leydens and Lucena, *Engineering Justice*.

19. Li, *Unearthing Conflict*; Smith and Smith, "Engineering and the Politics of Commensuration."

20. Delborne, Kokotovich, and Lunshof, "Social License and Synthetic Biology."

21. Social acceptance and social responsibility are different, though some engineers collapsed them, for example, by believing that one's social responsibilities were fulfilled as long as social acceptance was achieved. In this book I have teased out the differences between social acceptance and social responsibility principally by considering the question of whether engineers and the companies they work for are willing to consider the question not just how but whether resource extraction should take place.

22. For a description of the courses and activities, see Smith et al., "Student Learning about Engineering and Corporate Social Responsibility."

23. For more details on how the survey was developed and validated, see Smith et al., "Student Learning about Engineering and Corporate Social Responsibility."

24. Greg Rulifson, Shurraya Denning, Cassidy Grady, Juliana Lucena, Christopher Spotts, and Courtney Stanton were indispensable in helping to organize, clean up, and analyze thousands of undergraduate student survey responses.

25. Smith et al., "Industry–University Partnerships"; Smith, McClelland, and Smith, "Engineering Students' Views of Corporate Social Responsibility."

26. Smith et al., "Industry–University Partnerships."

27. Smith et al., "Student Learning about Engineering and Corporate Social Responsibility."

28. Smith et al., "Counteracting the Social Responsibility Slump?"

29. Smith et al., "Counteracting the Social Responsibility Slump?"

30. Canney and Bielefeldt, "Framework for the Development of Social Responsibility in Engineers."

31. The rest of this section draws from a co-authored ASEE conference paper noted in the acknowledgements to this book: Smith et al., "Counteracting the Social Responsibility Slump?"

32. Smith et al., "Counteracting the Social Responsibility Slump?"

33. Rajak, *In Good Company*, 11.

34. "Corporations shape human experience not only in spectacular and disastrous ways but also in mundane, everyday, ambivalent, and positive ways. They are, after all, the source of or conduit for much of what we wittingly and unwittingly produce and consume as we breathe, eat, drink, read, work, play, and move about the world. . . . No human alive today is breathing air or drinking water that has not been touched by corporate action." Welker, Partridge, and Hardin, "Corporate Lives," S4.

35. Downey and Lucena, "Engineering Selves," 120; Downey, "Engineering Cultures Syllabus as Formation Narrative," 455.

36. York, "Doing STS in STEM Spaces."

37. York, "Doing STS in STEM Spaces."

38. See chapter 7 of this book and Kelty, *Participant*.

Bibliography

PRIMARY SOURCES

Art Biddle personal archives, Golden, Colorado.

Crested Butte Mountain Heritage Museum, Crested Butte, Colorado.

Stan Dempsey personal archives, Golden, Colorado.

Mt. Emmons Mining Project collection, Leslie J. Savage Library, Western Colorado University, Gunnison, Colorado.

SECONDARY SOURCES

Abu-Lughod, Lila. "Can There Be a Feminist Ethnography?" *Women and Performance: A Journal of Feminist Theory* 5, no. 1 (1990): 7–27.

Ahearn, Laura M. "Language and Agency." *Annual Review of Anthropology* 30 (2001): 109–137.

Alder, Ken. *Engineering the Revolution: Arms and Enlightenment in France, 1763–1815.* Chicago: University of Chicago Press, 1997.

Anand, Nikhil, Akhil Gupta, and Hannah Appel. "Introduction: Temporality, Politics, and the Promise of Infrastructure. In *The Promise of Infrastructure*, edited by Nikhil Anand, Akhil Gupta, and Hannah Appel, 1–38. Durham, NC: Duke University Press Books, 2018.

Anand, Nikhil, Akhil Gupta, and Hannah Appel, eds. *The Promise of Infrastructure*. Durham, NC: Duke University Press, 2018.

Appadurai, Arjun. "Mediants, Materiality, Normativity." *Public Culture* 27, no. 2 (2015): 221–237. https://doi.org/10.1215/08992363-2841832.

Appel, Hannah. "Conclusion: Energy Ethics and Ethical Worlds." *Journal of the Royal Anthropological Institute* 25, no. S1 (2019): 177–190. https://doi.org/10.1111/1467-9655.13021.

Appel, Hannah. *The Licit Life of Capitalism: US Oil in Equatorial Guinea.* Durham, NC: Duke University Press, 2019.

Appel, Hannah. "Offshore Work: Oil, Modularity, and the How of Capitalism in Equatorial Guinea." *American Ethnologist* 39, no. 4 (2012): 692–709. https://doi.org/10.1111/j.1548-1425 .2012.01389.x.

Appel, Hannah. "To Critique or Not to Critique? That Is (Perhaps Not) the Question. . . ." *Journal of Business Anthropology* 8, no. 1 (2019): 1–6.

Appel, Hannah. "Walls and White Elephants: Oil Extraction, Responsibility, and Infrastructural Violence in Equatorial Guinea." *Ethnography* 13, no. 4 (2012): 439–465. https://doi.org /10.1177/1466138111435741.

Argenti, Nicolas, and Daniel M. Knight. "Sun, Wind, and the Rebirth of Extractive Economies: Renewable Energy Investment and Metanarratives of Crisis in Greece." *Journal of the Royal Anthropological Institute* 21, no. 4 (2015): 781–802. https://doi.org/10.1111/1467-9655.12287.

Auld, Graeme, Steven Bernstein, and Benjamin Cashore. "The New Corporate Social Responsibility." *Annual Review of Environment and Resources* 33 (2008): 413–435.

Bakker, Karen, and Gavin Bridge. "Material Worlds? Resource Geographies and the 'Matter of Nature.'" *Progress in Human Geography* 30, no. 1 (2006): 5–27.

Ballestero, Andrea. *A Future History of Water.* Durham, NC: Duke University Press Books, 2019.

Barandiarán, Javiera. *Science and Environment in Chile: The Politics of Expert Advice in a Neoliberal Democracy.* Cambridge, MA: MIT Press, 2018.

Barley, Stephen R., and Gideon Kunda. *Gurus, Hired Guns, and Warm Bodies: Itinerant Experts in a Knowledge Economy.* Princeton, NJ: Princeton University Press, 2006.

Barnes, Jessica. *Cultivating the Nile: The Everyday Politics of Water in Egypt.* Durham, NC: Duke University Press, 2014.

Barry, Andrew. *Material Politics: Disputes along the Pipeline.* Malden, MA: Wiley-Blackwell, 2013.

Benson, Peter. "El Campo: Faciality and Structural Violence in Farm Labor Camps." *Cultural Anthropology* 23, no. 4 (2008): 589–629.

Benson, Peter, and Stuart Kirsch. "Capitalism and the Politics of Resignation." *Current Anthropology* 51, no. 4 (2010): 459–486.

Berlant, Lauren, and Kathleen Stewart. *The Hundreds.* Durham, NC: Duke University Press, 2019.

Berry, Evan. *Devoted to Nature: The Religious Roots of American Environmentalism.* Berkeley: University of California Press, 2015.

Biddle, Art, Richard C. Livermore, and Adam Poe. "AMAX Inc and the Colorado Joint Review Process." New York: Center for Public Resources, September 1981.

Biersack, Aletta. " Reimagining Political Ecology: Culture/Power/History/Nature." In *Reimagining Political Ecology,* edited by Aletta Biersack and James B. Greenberg, 3–42. Durham, NC: Duke University Press, 2006.

Blowfield, Michael, and Jedrzej G. Frynas. "Setting New Agendas: Critical Perspectives on Corporate Social Responsibility in the Developing World." *International Affairs* 81, no. 3 (2005): 499–513.

Boas, Franz. "The Methods of Ethnology." *American Anthropologist* 22, no. 4 (1920): 311–321. https://doi.org/10.1525/aa.1920.22.4.02a00020.

Bourdieu, Pierre. *Outline of a Theory of Practice*. Cambridge: Cambridge University Press, 1977.

Boutilier, Robert, and Ian Thomson. *The Social License: The Story of the San Cristobal Mine*. New York: Routledge, 2018.

Bowker, Geoffrey. "Sustainable Knowledge Infrastructures." In *The Promise of Infrastructure*, edited by Nikhil Anand, Akhil Gupta, and Hannah Appel, 203–222. Durham, NC: Duke University Press, 2018.

Boyer, Dominic. "Infrastructure, Potential Energy, Revolution." In *The Promise of Infrastructure*, edited by Nikhil Anand, Akhil Gupta, and Hannah Appel, 223–244. Durham, NC: Duke University Press, 2018.

Boyer, Dominic. "Thinking through the Anthropology of Experts." *Anthropology in Action* 15, no. 2 (2008): 38–46. https://doi.org/10.3167/aia.2008.150204.

Bridge, Gavin, "Excavating Nature: Environmental Narratives and Discursive Regulation in the Mining Industry." In *An Unruly World? Globalization, Governance, and Geography*, edited by A. Herod, G. Tuathail, and S. Roberts, 219–243. New York: Routledge.

Browne, John. *Make, Think, Imagine: Engineering the Future of Civilization*. New York: Pegasus Books, 2019.

Brueckner, Martin, and Marian Eabrasu. "Pinning Down the Social License to Operate (SLO): The Problem of Normative Complexity." *Resources Policy* 59 (2018): 217–226. https://doi.org/10.1016/j.resourpol.2018.07.004.

Buch, Elana. "Senses of Care: Embodying Inequality and Sustaining Personhood in the Home Care of Older Adults in Chicago." *American Ethnologist* 40, no. 4 (2013): 637–650.

Butler, Judith. *Gender Trouble: Feminism and the Subversion of Identity*. New York: Routledge, 1990.

Cahn, Robert. *Footprints on the Planet: A Search for an Environmental Ethic*. New York: Universe Books, 1978.

Callon, Michel. "Some Elements of a Sociology of Translation: Domestication of the Scallops and the Fishermen of St Brieuc Bay." In *Power, Action, and Belief: A New Sociology of Knowledge*, edited by John Law, 196–233. London: Routledge and Kegan Paul, 1986.

Callon, Michel, Pierre Lascoumes, and Yannick Barthe, eds. *Acting in an Uncertain World*, translated by Graham Burchell. Cambridge, MA: MIT Press, 2011.

Canney, Nathan, and Angela Bielefeldt. "A Framework for the Development of Social Responsibility in Engineers." *International Journal of Engineering Education* 31 (2015): 414–424.

Carrier, James G., ed. *After the Crisis: Anthropological Thought, Neoliberalism, and the Aftermath.* New York: Routledge, 2016.

Carsten, Janet, Sophie Day, and Charles Stafford. "Introduction: Reason and Passion: The Parallel Worlds of Ethnography and Biography." *Social Anthropology* 26, no. 1 (2018): 5–14. https://doi.org/10.1111/1469-8676.12490.

Catalano, George D. "Engineering Ethics: Peace, Justice, and the Earth." *Synthesis Lectures on Engineers, Technology, and Society* 1, no. 1 (2006): 1–80. https://doi.org/10.2200/S00039ED1V01Y200606ETS001.

Cech, Erin A. "Culture of Disengagement in Engineering Education?" *Science, Technology, and Human Values* 39, no. 1 (2014): 42–72. https://doi.org/10.1177/0162243913504305.

Certeau, Michel de. *The Practice of Everyday Life.* Berkeley: University of California Press, 1984.

Chakrabarty, Dipesh. "The Climate of History: Four Theses." *Critical Inquiry* 35, no. 2 (2009): 197–222. https://doi.org/10.1086/596640.

Chapman, Chelsea. "Multinatural Resources: Ontologies of Energy and the Politics of Inevitability in Alaska." In *Cultures of Energy: Power, Practices, Technologies*, edited by Sarah Strauss, Stephanie Rupp, and Thomas Love, 96–109. Walnut Creek, CA: Left Coast Press, 2013.

Cloud, Dana L. "Corporate Social Responsibility as Oxymoron: Universalization and Exploitation at Boeing." In *The Debate over Corporate Social Responsibility*, edited by Steve May, George Cheney, and Juliet Roper, 219–231. Oxford: Oxford University Press, 2007.

Colby, Anne, and William M. Sullivan. "Ethics Teaching in Undergraduate Engineering Education." *Journal of Engineering Education* 97, no. 3 (2008): 327–338.

Collier, Paul, and Glen Ireland. "Shared-Use Mining Infrastructure: Why It Matters and How to Achieve It." *Development Policy Review* 36, no. 1 (2018): 51–68. https://doi.org/10.1111/dpr.12231.

Conley, John M., and Cynthia A. Williams. "Engage, Embed, and Embellish: Theory versus Practice in the Corporate Social Responsibility Movement." *Journal of Corporation Law* 31, no. 1 (2005): 1–38.

Cronon, William. "The Trouble with Wilderness; Or, Getting Back to the Wrong Nature." In *Uncommon Ground: Toward Reinventing Nature*, 69–90. New York: Norton, 1995.

Cross, Jamie. "Detachment as a Corporate Ethic: Materializing CSR in the Diamond Supply Chain." *Focaal* 2011, no. 60 (2011): 34–46. https://doi.org/10.3167/fcl.2011.600104.

Cumming, Daniel G. "Black Gold, White Power: Mapping Oil, Real Estate, and Racial Segregation in the Los Angeles Basin, 1900–1939." *Engaging Science, Technology, and Society* 4 (2018): 85–110. https://doi.org/10.17351/ests2018.212.

Dahlgren, Kari. "Digging Deeper: Precarious Futures in Two Australian Coal Mining Towns." PhD diss., London School of Economics, 2019.

Davis, Michael. "Professional Autonomy: A Framework for Empirical Research." *Business Ethics Quarterly* 6, no. 4 (1996): 441–460.

Davis, Rachel and Daniel Franks. *Costs of Company-Community Conflict in the Extractive Sector. Corporate Social Responsibility Initiative Report.* Cambridge, MA: John F. Kennedy School of Government, Harvard University, 2014. Avaailable online http://www.hks.harvard.edu/m-rcbg /CSRI/research/Costs%20of%20Conflict_Davis%20%20Franks.pdf.

Delborne, Jason A., Adam E. Kokotovich, and Jeantine E. Lunshof. "Social License and Synthetic Biology: The Trouble with Mining Terms." *Journal of Responsible Innovation* 7, no. 3 (2020): 280–297. https://doi.org/10.1080/23299460.2020.1738023.

Dolan, Catherine, and Dinah Rajak. "Introduction: Ethnographies of Corporate Ethicizing." *Focaal* 60 (2011): 3–8.

Dougherty, Michael L. "Boom Times for Technocrats? How Environmental Consulting Companies Shape Mining Governance." *Extractive Industries and Society* 6, no. 2 (2019): 443–454. https://doi.org/10.1016/j.exis.2019.01.007.

Downey, Gary Lee. "Are Engineers Losing Control of Technology? From 'Problem Solving' to 'Problem Definition and Solution' in Engineering Education." *Chemical Engineering Research and Design* 83, no. 6 (2005): 583–595. https://doi.org/10.1205/cherd.05095.

Downey, Gary Lee. "Critical Participation: Inflecting Dominant Knowledge Practices through STS." In *Making and Doing: Activating STS through Knowledge Expression and Travel*, edited by Gary Lee Downey and Teun Zuiderent-Jerak, 219–243. Cambridge, MA: MIT Press, 2021.

Downey, Gary Lee. "The Engineering Cultures Syllabus as Formation Narrative: Critical Participation in Engineering Education through Problem Definition." *University of St. Thomas Law Journal* 5, no. 2 (2011): 428–456.

Downey, Gary Lee. "Low Cost, Mass Use: American Engineers and the Metrics of Progress." *History and Technology* 23, no. 3 (2007): 289–308. https://doi.org/10.1080/07341510701300387.

Downey, Gary Lee. *The Machine in Me: An Anthropologist Sits among Computer Engineers.* New York: Routledge, 1998.

Downey, Gary Lee. "PDS: Engineering as Problem Definition and Solution." In *International Perspectives on Engineering Education*, edited by Steen Hyldgaard Christensen, Christelle Didier, Andrew Jamison, Martin Meganck, Carl Mitcham, and Byron Newberry, 435–455. Cham: Springer, 2015. https://doi.org/10.1007/978-3-319-16169-3_21.

Downey, Gary Lee. "What Is Engineering Studies For? Dominant Practices and Scalable Scholarship." *Engineering Studies* 1, no. 1 (2009): 55–76. https://doi.org/10.1080/19378620902786499.

Downey, Gary, and Juan Lucena. "Engineering Selves: Hiring into a Contested Field of Education." In *Cyborgs and Citadels: Anthropological Interventions in Emerging Sciences and Technologies*, edited by Gary Lee Downey and Joseph Dumit, 117–142. Santa Fe, NM: School of American Research Press, 1997.

Eagleton, Terry. *Ideology: An Introduction.* New York: Verso, 1991.

Edwards, Richard C. *Contested Terrain: The Transformation of the Workplace in the Twentieth Century.* New York: Basic Books, 1979.

Engineering and Mining Journal. "Openness, Cooperation—Keys to the AMAX Environmental Program." September 1972, 1–5.

Espig, Martin, and Kim de Rijke. "Unconventional Gas Developments and the Politics of Risk and Knowledge in Australia." *Energy Research and Social Science*, 20 (2016): 82–90. https://doi .org/10.1016/j.erss.2016.06.001.

Faulkner, Wendy. "'Nuts and Bolts and People': Gender-Troubled Engineering Identities." *Social Studies of Science* 37, no. 3 (2007): 331–356.

Ferguson, James. *The Anti-politics Machine: Development, Depoliticization, and Bureaucratic Power in Lesotho.* Minneapolis: University of Minnesota Press, 1990.

Ferguson, James. *Global Shadows: Africa in the Neoliberal World Order.* Durham, NC: Duke University Press, 2006.

Ferry, Elizabeth Emma, and Mandana E. Limbert. *Timely Assets: The Politics of Resources and Their Temporalities.* Santa Fe, NM: School for Advanced Research Press, 2008.

Folch, Christine. *Hydropolitics: The Itaipu Dam, Sovereignty, and the Engineering of Modern South America.* Princeton, NJ: Princeton University Press, 2019.

Forsythe, Diana. *Studying Those Who Study Us: An Anthropologist in the World of Artificial Intelligence,* edited by David Hess. Palo Alto, CA: Stanford University Press, 2002.

Fortun, Kim. *Advocacy after Bhopal: Environmentalism, Disaster, New Global Orders.* Chicago: University of Chicago Press, 2001.

Foucault, Michel. *The History of Sexuality.* Vol. 1. New York: Vintage, 1978.

Franks, Daniel. *Mountain Movers: Mining, Sustainability, and the Agents of Change.* New York: Routledge, 2015.

Franks, Daniel M., Rachel Davis, Anthony J. Bebbington, Saleem H. Ali, Deanna Kemp, and Martin Scurrah. "Conflict Translates Environmental and Social Risk into Business Costs." *Proceedings of the National Academy of Sciences* 111, no. 21 (2014): 7576–7581. https://doi.org/10 .1073/pnas.1405135111.

Frynas, Jedrzej George. "Political Instability and Business: Focus on Shell in Nigeria." *Third World Quarterly* 19, no. 3 (1998): 457–478.

Funtowicz, Silvio O., and Jerome R. Ravetz, "Science for the Post-normal Age." *Futures* 25, no. 7 (1993): 739–755. https://doi.org/10.1016/0016-3287(93)90022-L.

Gardner, Katy. *Discordant Development: Global Capitalism and the Struggle for Connection in Bangladesh.* London: Pluto Press, 2012.

Gardner, Katy, Zahir Ahmed, Fatema Bashir, and Masud Rana. "Elusive Partnerships: Gas Extraction and CSR in Bangladesh." *Resources Policy* 37, no. 2 (2012): 168–174. https://doi.org /10.1016/j.resourpol.2012.01.001.

Geertz, Clifford. "Thick Description: Toward an Interpretive Theory of Culture." In *The Interpretation of Cultures: Selected Essays,* 3–30. New York: Basic Books, 1973.

Gehman, Joel, Lianne M. Lefsrud, and Stewart Fast. "Social License to Operate: Legitimacy by Another Name?" *Canadian Public Administration* 60, no. 2 (2017): 293–317. https://doi.org/10.1111/capa.12218.

Giddens, Anthony. *Central Problems in Social Theory: Action, Structure and Contradiction in Social Analysis*. Berkeley: University of California Press, 1979.

Gilbert, Paul Robert, and Jessica Sklair. "Introduction: Ethnographic Engagements with Global Elites." *Focaal* 2018, no. 81 (2018): 1–15. https://doi.org/10.3167/fcl.2018.810101.

Gilberthorpe, Emma, and Glenn Banks. "Development on Whose Terms? CSR Discourse and Social Realities in Papua New Guinea's Extractive Industries Sector." *Resources Policy* 37, no. 2 (2012): 185–193. https://doi.org/10.1016/j.resourpol.2011.09.005.

Goldman, Steven L. "Why We Need a Philosophy of Engineering: A Work in Progress." *Interdisciplinary Science Reviews* 29, no. 2 (2004): 163–176. https://doi.org/10.1179/030801804225012572.

Gottlieb, Robert. *Forcing the Spring: The Transformation of the American Environmental Movement*. Rev. ed. Washington, DC: Island Press, 2005.

Graeber, David. *Toward an Anthropological Theory of Value: The False Coin of Our Own Dreams*. New York: Palgrave, 2001.

Gusterson, Hugh. *Nuclear Rites: A Weapons Laboratory at the End of the Cold War*. Berkeley: University of California Press, 1996.

Haggerty, Julia H., Adrianne C. Kroepsch, Kathryn Bills Walsh, Kristin K. Smith, and David W. Bowen. "Geographies of Impact and the Impacts of Geography: Unconventional Oil and Gas in the American West." *Extractive Industries and Society* 5, no. 4 (2018): 619–633. https://doi.org/10.1016/j.exis.2018.07.002.

Haraway, Donna. *Modest Witness@Second Millennium. FemaleMan Meets OncoMouse: Feminism and Technoscience*. New York: Routledge, 1997.

Harvey, Penny, and Hannah Knox. *Roads: An Anthropology of Infrastructure and Expertise*. Ithaca, NY: Cornell University Press, 2015.

Harvey, Penelope, and Hannah Knox. "Virtuous Detachments in Engineering Practice—On the Ethics of (Not) Making a Difference." In *Detachment: Essays on the Limits of Relational Thinking*, edited by Thomas Yarrow, Matei Candea, Catherine Trundle, and Joanna Cook, 58–78. Manchester: Manchester University Press, 2015.

Hawken, Paul, Amory Lovins, and L. Hunter Lovins. *Natural Capitalism: Creating the Next Industrial Revolution*. New York: US Green Building Council, 2000.

Hébert, Karen. "Chronicle of a Disaster Foretold: Scientific Risk Assessment, Public Participation, and the Politics of Imperilment in Bristol Bay, Alaska." *Journal of the Royal Anthropological Institute* 22, no. S1 (2016): 108–126. https://doi.org/10.1111/1467-9655.12396.

Herkert, Joseph R. "Future Directions in Engineering Ethics Research: Microethics, Macroethics, and the Role of Professional Societies." *Science and Engineering Ethics* 7, no. 3 (2001): 403–414. https://doi.org/10.1007/s11948-001-0062-2.

High, Mette M. *Fear and Fortune: Spirit Worlds and Emerging Economies in the Mongolian Gold Rush*. Ithaca, NY: Cornell University Press.

High, Mette M. "Projects of Devotion: Energy Exploration and Moral Ambition in the Cosmoeconomy of Oil and Gas in the Western United States." *Journal of the Royal Anthropological Institute* 25, no. S1 (2019): 29–46. https://doi.org/10.1111/1467-9655.13013.

High, Mette M., and Jessica M. Smith. "Introduction: The Ethical Constitution of Energy Dilemmas." *Journal of the Royal Anthropological Institute* 25, no. S1 (2019): 9–28. https://doi.org/10.1111/1467-9655.13012.

Ho, Karen. *Liquidated: An Ethnography of Wall Street*. Durham, NC: Duke University Press, 2009.

Hochschild, Arlie Russell. *Strangers in Their Own Land: Anger and Mourning on the American Right*. New York: New Press, 2016.

Hoffman, Andrew. *From Heresy to Dogma: An Institutional History of Corporate Environmentalism*. Expanded ed. Palo Alto, CA: Stanford University Press, 2001.

Hommels, Anique. *Unbuilding Cities: Obduracy in Urban Sociotechnical Change*. Cambridge, MA: MIT Press, 2008.

Horowitz, Leah S., and Michael Watts, eds. *Grassroots Environmental Governance: Community Engagements with Industry*. New York: Routledge, 2016.

Howe, Cymene. *Ecologics: Wind and Power in the Anthropocene*. Durham, NC: Duke University Press, 2019.

Huber, Matthew T. *Lifeblood: Oil, Freedom, and the Forces of Capital*. Minneapolis: University of Minnesota Press, 2013.

Hughes, David McDermott. *Energy without Conscience: Oil, Climate Change, and Complicity*. Durham, NC: Duke University Press Books, 2017.

Jacka, Jerry K. *Alchemy in the Rain Forest: Politics, Ecology, and Resilience in a New Guinea Mining Area*. Durham, NC: Duke University Press, 2015.

Jacka, Jerry K. "The Anthropology of Mining: The Social and Environmental Impacts of Resource Extraction in the Mineral Age." *Annual Review of Anthropology* 47 (2018): 61–77.

Jacquet, Jeffrey B. "Review of Risks to Communities from Shale Energy Development." *Environmental Science and Technology* 48, no. 15 (2014): 8321–8333. https://doi.org/10.1021/es404647x.

Jalbert, Kirk, Abby J. Kinchy, and Simona L. Perry. "Civil Society Research and Marcellus Shale Natural Gas Development: Results of a Survey of Volunteer Water Monitoring Organizations." *Journal of Environmental Studies and Sciences* 4, no. 1 (2014): 78–86. https://doi.org/10.1007/s13412-013-0155-7.

Jenkins, Philip. *Decade of Nightmares: The End of the Sixties and the Making of Eighties America*. New York: Oxford University Press, 2008.

Jobson, Ryan Cecil. "Dead Labor: On Racial Capital and Fossil Capital." In *Histories of Racial Capitalism*, edited by Destin Jenkins and Justin Leroy, 215–230. New York: Columbia University Press, 2021.

Johnson, Deborah. "Rethinking the Social Responsibilities of Engineers as a Form of Accountability." In *Philosophy and Engineering: Exploring Boundaries, Expanding Connections*, edited by Diane P. Michelfelder, Brian Newberry, and Qin Zhu, 85–98. New York: Springer, 2017.

Journal of Metals. "The Henderson Mine/Mill/Concentrator: Fifty Million Pounds of Molybdenum per Year from a Successful Experiment in Ecology." October 1977, 9–12.

Kaiser, David. *How the Hippies Saved Physics: Science, Counterculture, and the Quantum Revival*. New York: Norton, 2012.

Kaiser, David, and W. Patrick McCray, eds. *Groovy Science: Knowledge, Innovation, and American Counterculture*. Chicago: University of Chicago Press, 2016.

Karwat, Darshan M. A. "Self-Reflection for Activist Engineering." *Science and Engineering Ethics* 26, no. 3 (2020): 1329–1352. https://doi.org/10.1007/s11948-019-00150-y.

Kelty, Christopher. *The Participant: A Century of Participation in Four Stories*. Chicago: University of Chicago Press, 2020.

Kelty, Christopher. "Responsibility: McKeon and Ricoeur." Anthropology of the Contemporary Research Collaboratory, working paper no. 12, May 2008. http://evols.library.manoa.hawaii.edu/handle/10524/1625.

Kinchy, Abby. "Citizen Science and Democracy: Participatory Water Monitoring in the Marcellus Shale Fracking Boom." *Science as Culture* 26, no. 1 (2017): 88–110. https://doi.org/10.1080/09505431.2016.1223113.

Kirsch, Stuart. *Mining Capitalism: The Relationship between Corporations and Their Critics*. Berkeley, CA: University of California Press, 2014.

Kirsch, Stuart. *Reverse Anthropology: Indigenous Analysis of Social and Environmental Relations in New Guinea*. Stanford, CA: Stanford University Press, 2006.

Kirsch, Stuart. "Sustainable Mining." *Dialectical Anthropology* 34 (2010): 87–93.

Kneas, David. "Emergence and Aftermath: The (Un)Becoming of Resources and Identities in Northwestern Ecuador." *American Anthropologist* 120, no. 4 (2018): 752–764. https://doi.org/10.1111/aman.13150.

Kneas, David. "From Dearth to El Dorado: Andean Nature, Plate Tectonics, and the Ontologies of Ecuadorian Resource Wealth." *Engaging Science, Technology, and Society* 4 (2018): 131–154. https://doi.org/10.17351/ests2018.214.

Kneas, David. "Subsoil Abundance and Surface Absence: A Junior Mining Company and Its Performance of Prognosis in Northwestern Ecuador." *Journal of the Royal Anthropological Institute* 22, no. S1 (2016): 67–86. https://doi.org/10.1111/1467-9655.12394.

Knight, Peter. *Profits and Principles—Does There Have to Be a Choice?* (The Shell Report). London: Shell International, 1998.

Kroepsch, Adrianne. "Horizontal Drilling, Changing Patterns of Extraction, and Piecemeal Participation: Urban Hydrocarbon Governance in Colorado." *Energy Policy* 120 (2018): 469–480. https://doi.org/10.1016/j.enpol.2018.04.074.

Kroepsch, Adrianne. "New Rig on the Block: Spatial Policy Discourse and the New Suburban Geography of Energy Production on Colorado's Front Range." *Environmental Communication* 10, no. 3 (2016): 337–351. https://doi.org/10.1080/17524032.2015.1127852.

Kuipers, J. R., et al. "Comparison of Predicted and Actual Water Quality at Hardrock Mines: The Reliability of Predictions in Environmental Impact Statements." Washington, DC: Earthworks, 2006.

Kunda, Gideon. *Engineering Culture: Control and Commitment in a High-Tech Corporation.* Rev. ed. Philadelphia: Temple University Press, 2006.

Kusserow, Adrie Suzanne. "De-homogenizing American Individualism: Socializing Hard and Soft Individualism in Manhattan and Queens." *Ethos* 27, no. 2 (1999): 210–234.

Laidlaw, James. "Ethics." In *A Companion to the Anthropology of Religion*, edited by Janice Boddy and Michael Lambek, 169–188. Chichester, UK: Wiley, 2013.

Lamb, Sarah. "The Making and Unmaking of Persons: Notes on Aging and Gender in North India." *Ethos* 25, no. 3 (1997): 279–302.

Latour, Bruno. *Reassembling the Social: An Introduction to Actor-Network Theory.* Oxford: Oxford University Press, 2005.

Latour, Bruno. *We Have Never Been Modern*, translated by Catherine Porter. Cambridge, MA: Harvard University Press, 1993.

Latour, Bruno. "Why Has Critique Run Out of Steam? From Matters of Fact to Matters of Concern." *Critical Inquiry* 30, no. 2 (2004): 225–248. https://doi.org/10.1086/421123.

Layton, Edwin T., Jr. *The Revolt of the Engineers: Social Responsibility and the American Engineering Profession.* Baltimore: Johns Hopkins University Press, 1986.

LeCain, Timothy J. *Mass Destruction: The Men and Giant Mines That Wired America and Scarred the Planet.* New Brunswick, NJ: Rutgers University Press, 2009.

Leydens, Jon A., and Juan C. Lucena. *Engineering Justice: Transforming Engineering Education and Practice.* Hoboken, NJ: Wiley-IEEE Press, 2017.

Li, Fabiana. "Documenting Accountability: Environmental Impact Assessment in a Peruvian Mining Project." *PoLAR: Political and Legal Anthropology Review* 32, no. 2 (2009): 218–236.

Li, Fabiana. "Engineering Responsibility: Environmental Mitigation and the Limits of Commensuration in a Chilean Mining Project." *Focaal* 60 (2011): 61–73.

Li, Fabiana. *Unearthing Conflict: Corporate Mining, Activism, and Expertise in Peru.* Durham, NC: Duke University Press, 2015.

Livesey, Sharon M., and Kate Kearins. "Transparent and Caring Corporations? A Study of Sustainability Reports by the Body Shop and Royal Dutch/Shell." *Organization and Environment* 15, no. 3 (2002): 233–258.

Love, Thomas, and Anna Garwood. "Electrifying Transitions: Power and Culture in Rural Cajamarca, Peru." In *Cultures of Energy: Power, Practices, Technologies*, edited by Sarah Strauss, Stephanie Rupp, and Thomas Love, 147–163. Walnut Creek, CA: Left Coast Press, 2013.

Lucena, Juan C. *Defending the Nation: U.S. Policymaking to Create Scientists and Engineers from Sputnik to the "War against Terrorism."* Lanham, MA: University Press of America, 2005.

Lucena, Juan C. "Engineers and Community: How Sustainable Engineering Depends on Engineers' Views of People." In *Handbook of Sustainable Engineering*, edited by Joanne Kauffman and Kun-Mo Lee, 793–815. Dordrecht: Springer Netherlands, 2013. https://doi.org/10.1007/978-1-4020-8939-8_51.

Lucena, Juan, Jen Schneider, and Jon A. Leydens. "Engineering and Sustainable Community Development." *Synthesis Lectures on Engineers, Technology, and Society* 5, no. 1 (2010): 1–230.

Mahmood, Saba. *Politics of Piety: The Islamic Revival and the Feminist Subject*. Princeton, NJ: Princeton University Press, 2005.

Marchand, Roland. *Creating the Corporate Soul: The Rise of Public Relations and Corporate Imagery in American Big Business*. Berkeley: University of California Press, 1998.

Mathews, Andrew S., and Jessica Barnes. "Prognosis: Visions of Environmental Futures." *Journal of the Royal Anthropological Institute* 22, no. S1 (2016): 9–26. https://doi.org/10.1111/1467-9655.12391.

Mauss, Marcel. "A Category of the Human Mind: The Notion of Person, the Notion of Self." In *The Category of the Person: Anthropology, Philosophy, History*, edited by Michael Carrithers, Steven Collins, and Steven Lukes, 1–23. London: Routledge and Kegan Paul, 1979.

Mcintosh, Janet. "Personhood, Self, and Individual." In *The International Encyclopedia of Anthropology*, 4583–4591. American Cancer Society, 2018. https://doi.org/10.1002/9781118924396.wbiea1576.

McNeil, Bryan T. *Combating Mountaintop Removal: New Directions in the Fight against Big Coal*. Urbana: University of Illinois Press, 2011.

Meiksins, Peter. "The 'Revolt of the Engineers' Reconsidered." *Technology and Culture* 29, no. 2 (1988): 219–246. https://doi.org/10.2307/3105524.

Meiksins, Peter, and Chris Smith. "Why American Engineers Aren't Unionized: A Comparative Perspective." *Theory and Society* 22, no. 1 (1993): 57–97. https://doi.org/10.1007/BF00993448.

Meiksins, Peter F., and James M. Watson. "Professional Autonomy and Organizational Constraint: The Case of Engineers." *Sociological Quarterly* 30, no. 4 (1989): 561–685.

Meiksins, Peter, and Peter Whalley. *Putting Work in Its Place: A Quiet Revolution*. Ithaca, NY: Cornell University Press, 2002.

Metze, Tamara. "Framing the Future of Fracking: Discursive Lock-in or Energy Degrowth in the Netherlands?" *Journal of Cleaner Production* 197 (2018): 1737–1745. https://doi.org/10.1016/j.jclepro.2017.04.158.

Miller, George A. "Professionals in Bureaucracy: Alienation among Industrial Scientists and Engineers." *American Sociological Review* 32, no. 5 (1967): 755–768. https://doi.org/10.2307/2092023.

Miller, Gloria E. "Frontier Masculinity in the Oil Industry: The Experience of Women Engineers." *Gender, Work, and Organization* 11, no. 1 (2004): 47–73.

Minteer, Ben. "Pragmatism, Piety, and Environmental Ethics." *Worldviews: Global Religions, Culture, and Ecology* 12, no. 2–3 (2008): 179–196. https://doi.org/10.1163/156853508X359976.

Mitcham, Carl. "A Historico-Ethical Perspective on Engineering Education: From Use and Convenience to Policy Engagement." *Engineering Studies* 1, no. 1 (2009): 35–53. https://doi.org/10.1080/19378620902725166.

Mitcham, Carl. *Steps toward a Philosophy of Engineering*. Lanham, MD: Rowman and Littlefield, 2019.

Mitcham, Carl. "The True Grand Challenge for Engineering: Self-Knowledge." *Issues in Science and Technology* 31, no. 1 (2014). https://issues.org/perspectives-the-true-grand-challenge-for-engineering-self-knowledge/.

Mitcham, Carl, and Nan Wang. "From Engineering Ethics to Engineering Politics." In *Engineering Identities, Epistemologies, and Values*, edited by Steen Hyldgaard Christensen, Christelle Didier, Andrew Jamison, Martin Meganck, Carl Mitcham, and Byron Newberry, 307–324. Cham: Springer, 2015. https://doi.org/10.1007/978-3-319-16172-3_17.

Mitchell, Timothy. *Rule of Experts: Egypt, Techno-politics, Modernity*. Berkeley: University of California Press, 2002.

Mody, Cyrus. *The Squares: US Physical and Engineering Scientists in the Long 1970s*. Cambridge, MA: MIT Press, 2021.

Mol, Annemarie. *The Body Multiple: Ontology in Medical Practice*. Durham, NC: Duke University Press, 2002.

Morrison, John. *The Social License: How to Keep Your Organization Legitimate*. London: Palgrave Macmillan, 2014.

Müftüoglu, Ingrid Birce, Ståle Knudsen, Ragnhild Freng Dale, Oda Eiken, Dinah Rajak, and Siri Lange. "Rethinking Access: Key Methodological Challenges in Studying Energy Companies." *Energy Research and Social Science* 45 (2018): 250–257. https://doi.org/10.1016/j.erss.2018.07.019.

Munoz, Jose-Maria, and Philip Burnham. "Subcontracting as Corporate Social Responsibility in the Chad-Cameroon Pipeline Project." In *The Anthropology of Corporate Social Responsibility*, edited by Catherine Dolan and Dinah Rajak, 152–178. Oxford: Berghahn Books, 2016.

Murphy, Patrick E. "Corporate Social Responsiveness: An Evolution." *University of Michigan Business Review* 30, no. 6 (1978): 19–25, 30.

Nader, Laura. "Controlling Processes: Tracing the Dynamic Components of Power." *Current Anthropology* 38, no. 5 (1997): 711–737.

Nieusma, Dean. "Analyzing Context by Design: Engineering Education Reform via Social-Technical Integration." In *International Perspectives on Engineering Education*, edited by S.

Christensen, C. Didier, A. Jamison, M. Meganck, C. Mitcham, and B. Newberry, 415–434. Cham: Springer, 2015. https://doi.org/10.1007/978-3-319-16169-3_20.

Nieusma, Dean. "Conducting the Instrumentalists: A Framework for Engineering Liberal Education." *Engineering Studies* 7, no. 2–3 (2015): 159–163. https://doi.org/10.1080/19378629.2015.1085060.

Nixon, Rob. *Slow Violence and the Environmentalism of the Poor*. Cambridge, MA: Harvard University Press, 2011.

Noble, David F. *America by Design: Science, Technology, and the Rise of Corporate Capitalism*. Oxford: Oxford University Press, 1977.

Norman, W., and C. MacDonald. "Getting to the Bottom of 'Triple Bottom Line.'" *Business Ethics Quarterly* 14, no. 2: (2004): 243–262.

Ortner, Sherry. "Resistance and the Problem of Ethnographic Refusal." *Comparative Study of Society and History* 37, no. 1 (1995): 173–193.

Ortner, Sherry. "Theory in Anthropology since the Sixties." *Comparative Studies in Society and History* 26 (1984): 126–166.

Ottinger, Gwen. *Refining Expertise How Responsible Engineers Subvert Environmental Justice Challenges*. New York: New York University Press, 2013.

Owen, John. "Social License and the Fear of Mineras Interruptus." *Geoforum* 77 (2016): 102–105. https://doi.org/10.1016/j.geoforum.2016.10.014.

Owen, John R., and Deanna Kemp. *Extractive Relations: Countervailing Power and the Global Mining Industry*. New York: Routledge, 2017.

Owen, John R., and Deanna Kemp. "Social Licence and Mining: A Critical Perspective." *Resources Policy* 38, no. 1 (2013): 29–35. https://doi.org/10.1016/j.resourpol.2012.06.016.

Özden-Schilling, Canay. "The Infrastructure of Markets: From Electric Power to Electronic Data: Infrastructure of Markets." *Economic Anthropology* 3, no. 1 (2016): 68–80. https://doi.org/10.1002/sea2.12045.

Paxton, Jonijane. "Experiment in Ecology: AMAX/Henderson Meeting the Environmental Challenges." *Editorial Alert* (1974): 1–8.

Phadke, Roopali. "Green Energy Futures: Responsible Mining on Minnesota's Iron Range." *Energy Research and Social Science* 35 (2018): 163–173. https://doi.org/10.1016/j.erss.2017.10.036.

Porter, Michael E., and Mark R. Kramer. "Creating Shared Value." *Harvard Business Review*, January-February 2011. https://hbr.org/2011/01/the-big-idea-creating-shared-value.

Priest, Tyler. *The Offshore Imperative: Shell Oil's Search for Petroleum in Postwar America*. College Station: Texas A&M University Press, 2009.

Prno, Jason, and D. Scott Slocombe. "Exploring the Origins of 'Social License to Operate' in the Mining Sector: Perspectives from Governance and Sustainability Theories." *Resources Policy* 37, no. 3 (2012): 346–357. https://doi.org/10.1016/j.resourpol.2012.04.002.

Puig de la Bellacasa, María. "Ethical Doings in Naturecultures." *Ethics, Place, and Environment* 13, no. 2 (2010): 151–169. https://doi.org/10.1080/13668791003778834.

Rajak, Dinah. *In Good Company: An Anatomy of Corporate Social Responsibility*. Palo Alto, CA: Stanford University Press, 2011.

Rajak, Dinah. "Theatres of Virtue: Collaboration, Consensus, and the Social Life of Corporate Social Responsibility." *Focaal* 60 (2011): 9–20.

Reddy, Elizabeth. "Measuring Like an Engineer: Engineers Entangle Seismicity, Technologies, and Professional Identities in Mexico." Paper presented at the American Anthropological Association annual conference, Vancouver, 2019.

Richardson, Tanya, and Gisa Weszkalnys. "Introduction: Resource Materialities." *Anthropological Quarterly* 87, no. 1 (2014): 5–30. https://doi.org/10.1353/anq.2014.0007.

Riles, Annelise. "Too Big to Fail." In *Recasting Anthropological Knowledge: Inspiration and Social Science*, edited by Jeanette Edwards and Maja Petrovic-Steger, 31–48. Cambridge: Cambridge University Press, 2011.

Riley, Donna. "Engineering and Social Justice." *Synthesis Lectures on Engineers, Technology, and Society* 3, no. 1 (2008): 1–152. https://doi.org/10.2200/S00117ED1V01Y200805ETS007.

Rogers, Douglas. *The Depths of Russia: Oil, Power, and Culture after Socialism*. Ithaca, NY: Cornell University Press, 2015.

Rogers, Douglas. "The Materiality of the Corporation: Oil, Gas, and Corporate Social Technologies in the Remaking of a Russian Region." *American Ethnologist* 39, no. 2 (2012): 284–296.

Rolston, Jessica Smith. *Mining Coal and Undermining Gender: Rhythms of Work and Family in the American West*. New Brunswick, NJ: Rutgers University Press, 2014.

Rolston, Jessica Smith. "Turning Protesters into Monitors: Appraising Critical Collaboration in the Mining Industry." *Society and Natural Resources* 28, no. 2 (2015): 165–179. https://doi.org/10.1080/08941920.2014.945063.

Rose, Nikolas. *The Politics of Life Itself: Biomedicine, Power, and Subjectivity in the Twenty-First Century*. Princeton, NJ: Princeton University Press, 2007.

Rulifson, Greg, and Angela R. Bielefeldt. "Evolution of Students' Varied Conceptualizations about Socially Responsible Engineering: A Four Year Longitudinal Study." *Science and Engineering Ethics* 25 (2019): 939–974. https://doi.org/10.1007/s11948-018-0042-4.

Rulifson, Greg, and Angela Bielefeldt. "Motivations to Leave Engineering: Through a Lens of Social Responsibility." *Engineering Studies* 9, no. 3 (2017): 222–248. https://doi.org/10.1080/19378629.2017.1397159.

Sawyer, Suzana. *Crude Chronicles: Indigenous Politics, Multinational Oil, and Neoliberalism in Ecuador*. Durham, NC: Duke University Press, 2004.

Scheper-Hughes, Nancy. "Ire in Ireland." *Ethnography* 1, no. 1 (2000): 117–140.

Scott, James. *Seeing like a State: How Certain Schemes to Improve the Human Condition Have Failed*. New Haven, CT: Yale University Press, 1998.

Seely, Bruce. "Research, Engineering, and Science in American Engineering Colleges: 1900–1960." *Technology and Culture* 34, no. 2 (1993): 344–386. https://doi.org/10.2307/3106540.

Shever, Elana. "Engendering the Company: Corporate Personhood and the 'Face' of an Oil Company in Metropolitan Buenos Aires." *PoLAR: Political and Legal Anthropology Review* 33, no. 1 (2010): 26–46. https://doi.org/10.1111/j.1555-2934.2010.01091.x.

Shotwell, Alexis. *Against Purity: Living Ethically in Compromised Times*: Minneapolis: University of Minnesota Press, 2016.

Smith, Jessica M. "Boom to Bust, Ashes to (Coal) Dust: The Contested Ethics of Energy Exchanges in a Declining US Coal Market." *Journal of the Royal Anthropological Institute* 25, no. S1 (2019): 91–107. https://doi.org/10.1111/1467-9655.13016.

Smith, Jessica M. "The Ethics of Material Provisioning: Insiders' Views of Work in the Extractive Industries." *Extractive Industries and Society* 6, no. 3 (2019): 807–814. https://doi.org/10.1016/j.exis.2019.05.014.

Smith, Jessica M. "From Corporate Social Responsibility to Creating Shared Value: Contesting Responsibilization and the Mining Industry." In *Competing Responsibilities: The Ethics and Politics of Contemporary Life*, edited by Susanna Trnka and Catherine Trundle, 118–132. Durham, NC: Duke University Press, 2018.

Smith, Jessica, and Federico Helfgott. "Flexibility or Exploitation? Corporate Social Responsibility and the Perils of Universalization." *Anthropology Today* 26, no. 3 (2010): 20–23.

Smith, Jessica M., and Juan C. Lucena. "Socially Responsible Engineering." In *Routledge Handbook of the Philosophy of Engineering*, edited by Diane P. Michelfelder and Neelke Doorn, 661–673. New York: Routledge, 2021.

Smith, Jessica M., Carrie J. McClelland, and Nicole M. Smith. "Engineering Students' Views of Corporate Social Responsibility: A Case Study from Petroleum Engineering." *Science and Engineering Ethics* 23, no. 6 (2017): 1775–1790.

Smith, Jessica Mary, Greg Rulifson, Courtney Stanton, Nicole M. Smith, Linda A. Battalora, Emily Sarver, Carrie J. McClelland, Rennie B. Kaunda, and Elizabeth Holley. "Counteracting the Social Responsibility Slump? Assessing Changes in Student Knowledge and Attitudes in Mining, Petroleum, and Electrical Engineering," Paper presented at the American Society for Engineering Education conference (virtual), 2020.

Smith, Jessica M., and Nicole M. Smith. "Engineering and the Politics of Commensuration in the Mining and Petroleum Industries." *Engaging Science, Technology, and Society* 4 (2018): 67–84. https://doi.org/10.17351/ests2018.211.

Smith, Jessica Mary, Nicole M. Smith, Greg Rulifson, Carrie J. McClelland, Linda A. Battalora, Emily A. Sarver, and Rennie B. Kaunda. "Student Learning about Engineering and Corporate Social Responsibility: A Comparison across Engineering and Liberal Arts Courses." Paper presented at 2018 ASEE Annual Conference and Exposition, Salt Lake City, UT, 2018. https://peer.asee.org/student-learning-about-engineering-and-corporate-social-responsibility-a-comparison-across-engineering-and-liberal-arts-courses.

Smith, Jessica M., and Abraham S. D. Tidwell. "The Everyday Lives of Energy Transitions: Contested Sociotechnical Imaginaries in the American West." *Social Studies of Science* 46, no. 3 (2016): 327–350. https://doi.org/10.1177/0306312716644534.

Smith, Karl. "From Dividual and Individual Selves to Porous Subjects." *Australian Journal of Anthropology* 23, no. 1 (2012): 50–64. https://doi.org/10.1111/j.1757-6547.2012.00167.x.

Smith, Nicole M., Jessica M. Smith, Linda A. Battalora, and Benjamin A. Teschner. "Industry–University Partnerships: Engineering Education and Corporate Social Responsibility." *Journal of Professional Issues in Engineering Education and Practice* 144, no. 3 (2018): 04018002. https://doi .org/10.1061/(ASCE)EI.1943-5541.0000367.

Smith, Nicole, Jessica M. Smith, Qin Zhu, and Carl Mitcham. "Enhancing Engineering Ethics: Role Ethics and Corporate Social Responsibility." Manuscript forthcoming, *Science and Engineering Ethics,* 2020.

Star, Susan Leigh. "This Is Not a Boundary Object: Reflections on the Origin of a Concept." *Science, Technology, and Human Values* 35, no. 5 (2010): 601–617. https://doi.org/10.1177 /0162243910377624.

Star, Susan Leigh, and James R. Griesemer. "Institutional Ecology, 'Translations' and Boundary Objects: Amateurs and Professionals in Berkeley's Museum of Vertebrate Zoology, 1907–39." *Social Studies of Science* 19, no. 3 (1989): 387–420. https://doi.org/10.1177/030631289019003001.

Stein, Felix. *Work, Sleep, Repeat: The Abstract Labour of German Management Consultants.* London: Bloomsbury, 2019.

Strathern, Marilyn. *After Nature: English Kinship in the Late Twentieth Century.* Cambridge: Cambridge University Press, 1992.

Strathern, Marilyn. "Cutting the Network." *The Journal of the Royal Anthropological Institute* 2, no. 3 (1996): 517–535.

Strathern, Marilyn. *The Gender of the Gift: Problems with Women and Problems with Society in Melanesia.* Berkeley: University of California Press, 1988.

Strathern, Marilyn. "New Accountabilities: Anthropological Studies in Audit, Ethics and the Academy." In *Audit Culture: Anthropological Studies in Accountability, Ethics, and the Academy,* edited by Marilyn Strathern, 1–18. London: Routledge, 2000.

Tang, Xiofeng, and Dean Nieusma. "Contextualizing the Code: Ethical Support and Professional Interests in the Creation and Institutionalization of the 1974 IEEE Code of Ethics." *Engineering Studies* 9, no. 3 (2017): 166–194.

Thomson, Ian, and Robert Boutilier. "The Social Licence to Operate." In *SME Mining Engineering Handbook,* edited by Peter Darling, 1779–1796. Englewood, CO: Society for Mining, Metallurgy, and Exploration, 2011.

Trevelyan, James. *The Making of an Expert Engineer.* Boca Raton, FL: CRC Press, 2014.

Trnka, Susanna, and Catherine Trundle. "Introduction: Competing Responsibilities: Reckoning Personal Responsibility, Care for the Other, and the Social Contract in Contemporary Life."

Competing Responsibilities: The Ethics and Politics of Contemporary Life, edited by Susanna Trnka and Catherine Trundle, 1–25. Durham, NC: Duke University Press, 2017.

Watts, Michael. "Blood Oil: The Anatomy of a Petro-Insurgency in the Niger Delta." *Focaal* 2008, no. 52 (2008): 18–38. https://doi.org/10.3167/fcl.2008.520102.

Welker, Marina. "'Corporate Security Begins in the Community': Mining, the Corporate Social Responsibility Industry, and Environmental Advocacy in Indonesia." *Cultural Anthropology* 24, no. 1 (2009): 142–179.

Welker, Marina. *Enacting the Corporation: An American Mining Firm in Post-authoritarian Indonesia*. Berkeley: University of California Press, 2014.

Welker, Marina. "The Green Revolution's Ghost: Unruly Subjects of Participatory Development in Rural Indonesia." *American Ethnologist* 39, no. 2 (2012): 389–406.

Welker, Marina, Damani Partridge, and Rebecca Hardin. "Corporate Lives: New Perspectives on the Social Life of the Corporate Form; An Introduction to Supplement 3." *Current Anthropology* 52, no. S3 (2011): S3–S16.

Welker, Marina, and David Wood. "Shareholder Activism and Alienation." *Current Anthropology* 52, no. S3 (2011): S57–S69. https://doi.org/10.1086/656796.

Wellock, Thomas R. *Critical Masses: Opposition to Nuclear Power in California, 1958–1978*. Madison: University of Wisconsin Press, 1998.

Weszkalnys, Gisa. "Anticipating Oil: The Temporal Politics of a Disaster Yet to Come." *Sociological Review* 62 (2014): 211–335. https://doi.org/10.1111/1467-954X.12130.

Williams, Christine L., Kristine Kilanski, and Chandra Muller. "Corporate Diversity Programs and Gender Inequality in the Oil and Gas Industry." *Work and Occupations* 41, no. 4 (2014): 440–476. https://doi.org/10.1177/0730888414539172.

Williams, Christine L., Chandra Muller, and Kristine Kilanski. "Gendered Organizations in the New Economy." *Gender and Society* 26, no. 4 (2012): 549–573. https://doi.org/10.1177/0891243212445466.

Williamson, Bess. "Small-Scale Technology for the Developing World: Volunteers for International Technical Assistance, 1959–1971." *Comparative Technology Transfer and Society* 6, no. 3 (2008): 236–258. https://doi.org/10.1353/ctt.0.0019.

Willow, Anna J. *Understanding ExtrACTIVISM: Culture and Power in Natural Resource Disputes*. New York: Routledge, 2018.

Winner, Langdon. "Engineering Ethics and Political Imagination." In *Broad and Narrow Interpretations of Philosophy of Technology*, edited by Paul T. Durbin, 53–64. Dordrecht: Springer, 1990. https://doi.org/10.1007/978-94-009-0557-3_6.

Wisnioski, Matthew. *Engineers for Change: Competing Visions of Technology in 1960s America*. Cambridge, MA: MIT Press, 2012.

Wylie, Sara Ann. *Fractivism: Corporate Bodies and Chemical Bonds*. Durham, NC: Duke University Press, 2018.

Yarrow, Thomas. *Architects: Portraits of a Practice*. Ithaca, NY: Cornell University Press, 2019.

Yarrow, Thomas, Matei Candea, Catherine Trundle, and Joanna Cook. *Detachment: Essays on the Limits of Relational Thinking*. Oxford: Oxford University Press, 2015.

York, Emily. "Doing STS in STEM Spaces: Experiments in Critical Participation." *Engineering Studies* 10, no. 1 (2018): 66–84. https://doi.org/10.1080/19378629.2018.1447576.

Zhu, Qin. "Engineering Ethics Education, Ethical Leadership, and Confucian Ethics." *International Journal of Ethics Education* 3, no. 2 (2018): 169–179. https://doi.org/10.1007/s40889-018-0054-6.

Zhu, Qin, and Brent K. Jesiek. "A Pragmatic Approach to Ethical Decision-Making in Engineering Practice: Characteristics, Evaluation Criteria, and Implications for Instruction and Assessment." *Science and Engineering Ethics* 23, no. 3 (2017): 663–679. https://doi.org/10.1007/s11948-016-9826-6.

Zussman, Robert. *Mechanics of the Middle Class: Work and Politics among American Engineers*. Berkeley: University of California Press, 1985.

Index

Coal,
  controversy and, 64, 112, 239n23, 240n41, 248n25,
  ethic of material provisioning, 37
  labor and, 127, 238n18, 249n34
  production, 60
Codes of ethics, 197–198, 243n8, 244n21. *See also* Engineering ethics
  harmony ideologies, 35, 49–50, 56, 165
  loyalty to corporate employers, 33–35, 244n13
  pragmatism, 165
  professional idealism, 35
  technocratic efficiency, 34
Collaborations with engineering professors, 225–229
Coloradans for Responsible Energy Development (CRED), 164
Colorado. *See* Climax mine; Fracking controversies; Henderson mine; Mt. Emmons project
Colorado Joint Review Process (CJRP), 81–84, 90–95
Colorado Oil and Gas Association, 103
Colorado Open Space Coordinating Council, 67
Colorado School of Mines, 86, 129, 164, 219–222. *See also* Critical participation
  alumni, 21–22, 66, 71, 129, 192
  collaborations with engineering professors at, 225–229
  Corporate Social Responsibility course, 20–21, 49, 163, 223–225
  environmental education at, 69–70
  interdisciplinary teaching, 227–228
  participant observation at, xii, 19–20
  student knowledge and attitudes, 229–234
Commensuration, 120, 223
Community acceptance. *See* Social acceptance, business case for; Social license to operate (SLO)

Community engagement. *See* Public engagement strategies
Compensation, 180–183
Competing accountabilities, 6, 10, 23–32, 35. *See also* Engineering ethics; Ethic of material provisioning; Harmony ideologies; Social license to operate (SLO)
  attempted reconciliation of, 31–32, 125, 192–195, 215–217
  concept of, 27–28, 55–57
  consulting and, 136, 141–145, 159–160
  domains of accountability, 28–29, 192–193
  everyday practices of accountability, 28–32
  formal accountabilities, 28, 32–36, 192–193
  limitations of, 209–212
  moral architectures and, 41, 56–57, 164
  professional ideals, 33–36
Complicity, xv, 16, 37–38, 131, 140, 220, 235
Conferences, research value of, 21, 235236
Conflict, cost of. *See* Cost of conflict
Conley, John M., 95
ConocoPhillips, 102
Consultants, 24, 195
  alienation from implementation of work, 135–137, 145–149, 160–161
  alignment of projects with personal accountabilities, 130–131, 136, 141–145, 159–160
  compartmentalization by, 136, 231
  dependency on corporate forms, 137
  detachment from corporate forms, 137–140
  diversity and numbers of, 257n1
  narratives of reform, 149–155, 161–162
  participation in the corporate person, 139–140, 258n16
  political economy of, 161
  professional autonomy, 140–141, 159–160
  as sources of legitimacy, 155–159
Contextual listening, 167
Cooney, Jim, 42

Geological engineers
  as consultants, 135–136, 139, 150–154, 171
  detachment from corporate forms, 124–125, 135–136
  environmental ethics, 135–136, 151–152, 200–201
  ethical dilemmas, 31–32, 186–187, 198, 211, 228
  nonprofit work, 153
Gladwin, Thomas, 80, 251n54
Graeber, David, 255n32
Greenpeace, 59
Gunnison, Colorado. *See* Mt. Emmons project (Colorado)
Gusterson, Hugh, 20

Harmony ideologies, 49–55, 169, 193, 223. *See also* Competing accountabilities; Social license to operate (SLO); Win-win solutions
Harvard Kennedy School Corporate Social Responsibility Initiative, 43
Harvey, Penelope, 17, 138, 162, 244n12
Henderson Mine (Colorado), 66–71, 247n10
Hermeneutics of suspicion, 15–16, 193
High, Mette M., 17, 239n29
High Country Citizens Alliance, 78
Hoffman, Andrew, 63
Horizontal drilling, 1, 173, 262n39
Hughes, David McDermott, 14, 16, 242n97
Hydraulic fracturing. *See* Fracking controversies

Ideology of inevitability, 40, 90
Independent consultants. *See* Consultants
Independent Petroleum Association of America, 163
Individualization of blame, 8–9, 37, 205
Industrial environmentalism, 63
Informed consent. *See* Free, Prior, and Informed Consent

Infrastructure
  and engineers, 13, 17
  and publics, 241n78
  revolutionary, 214
  shared, 170–173
Institute of Electrical and Electronics Engineers (IEEE), 34–35
Interdisciplinary collaboration, xiv–xv, 225–229
International Council on Mining and Metals, 59, 61, 172
International Finance Corporation, 33, 181, 244n21
International Iron and Steel Institute, 74
International Nickel Company, 71
International Organization for Standardization (ISO), 244n21
International performance standards, 33, 50, 244n12, 244n21
Interviews, 242n87, 247n4. *See also* Research methodologies
  corporate self-representation revealed through, 19, 23, 109
  cultures of expertise and, 20–21
Izaak Walton League, 73

Jackson, D. C., 33, 244n15
Jobson, Ryan Cecil, 238n18
Johnson, Deborah, 30, 107

Kelty, Christopher, 62, 104, 204, 211
Kemp, Deanna, 15, 45, 206, 262n37
Kilby, Jack, 64
Kirsch, Stuart, 245n47, 246n67, 251n59
Kneas, David, 14
Knowledge workers, 213
Knox, Hannah, 17, 138, 162, 244n12
Kunda, Gideon, 106, 109, 116, 141

Labor, divisions of, 10, 99, 177, 195–197
Landmen, 21, 120